Praise for *The Big Fix*

"Full of illustrative stories and compelling evidence, *The Big Fix* outlines an ambitious yet feasible guide for addressing the climate crisis. Business leaders, activists, and policymakers at all levels will find inspiration from the pragmatic approaches outlined in this book."

—Al Gore, chairman of The Climate Reality Project, chairman of Generation Investment Management, and former vice president of the United States

"Nobody grasps climate policy—and what can actually work—better than Hal Harvey. Nobody elucidates climate science better than Justin Gillis. Together, they offer a bold blueprint for saving a habitable Earth."

—John Doerr, chairman of venture capital firm Kleiner Perkins and author of *Speed & Scale*

"A truly comprehensive—and entirely comprehensible—guide to the things we can and must do to transform our use of energy. This book will be of great use to anyone who wants to participate in the greatest technological revolution in human history."

—Bill McKibben, a founder of grassroots climate-campaign group 350.org and author of *The Flag, the Cross, and the Station Wagon*

"Smart, honest, and down-to-earth, *The Big Fix* addresses *the* crucial issue of our time: how citizens can compel action on climate change."

—**Elizabeth Kolbert, Pulitzer Prize-winning author of *The Sixth Extinction* and *Under a White Sky***

"[Harvey and] Gillis make fighting climate change feel a bit less intimidating in this down-to-earth look at ways the average citizen can make a difference . . . a useful guide for budding activists."

—***Publishers Weekly***

Also by HAL HARVEY

Designing Climate Solutions: A Policy Guide for Low-Carbon Energy
(with Robbie Orvis and Jeffrey Rissman)

Money Well Spent: A Strategic Plan for Smart Philanthropy
(with Paul Brest)

Security Without War: A Post–Cold War Foreign Policy
(with Michael H. Shuman)

The Big Fix

7 PRACTICAL STEPS TO SAVE OUR PLANET

HAL HARVEY and JUSTIN GILLIS

With MARK SILBERG and AMANDA MYERS

SIMON & SCHUSTER PAPERBACKS

New York London Toronto Sydney New Delhi

Simon & Schuster Paperbacks
An Imprint of Simon & Schuster, Inc.
1230 Avenue of the Americas
New York, NY 10020

Charts by Raaziq Masud Brown

First Simon & Schuster trade paperback edition June 2023

Interior design by Kyle Kabel

Manufactured in the United States of America

1 3 5 7 9 10 8 6 4 2

Library of Congress Cataloging-in-Publication Data

Names: Harvey, Hal, author | Gillis, Justin, author.
Title: The big fix : seven practical steps to save our planet / Hal Harvey & Justin Gillis.
Identifiers: LCCN 2021055164 (print) | LCCN 2021055165 (ebook) | ISBN 9781982123987 (hardcover) | ISBN 9781982124007 (ebook)
Subjects: LCSH: Climate change mitigation--Citizen participation. | Environmental policy--Citizen participation. | Energy policy--Citizen participation. | Sustainable development--Citizen participation. |
BISAC: POLITICAL SCIENCE / Public Policy / Environmental Policy / SCIENCE / Environmental Science (see also Chemistry / Environmental)
Classification: LCC TD171.75 .H375 2022 (print) | LCC TD171.75 (ebook) | DDC 363.738/746--dc23/eng/20220113
LC record available at https://lccn.loc.gov/2021055164
LC ebook record available at https://lccn.loc.gov/2021055165

ISBN 978-1-9821-2398-7
ISBN 978-1-9821-2399-4 (pbk)
ISBN 978-1-9821-2400-7 (ebook)

We dedicate this book to everyone who
is putting time, intelligence, and care into saving
our planet from the ravages of climate change

Contents

Introduction

The world is on fire.

The flames are hard to see, because we hide them so well. But you can hear them—in the whine of jet engines as planes streak across the sky, in the rumble of power plants as they send electricity surging over power lines, in the purr of your car engine as you drive to work.

When the American military pushed Saddam Hussein out of Kuwait in 1990, after his ill-advised invasion, his forces set fire to hundreds of oil wells, causing a hellish conflagration. Smoke reached all the way to Europe, and the fires could be seen from the International Space Station, their intensity prompting comparisons to Dante's Inferno. Ten thousand firefighters and other workers spent nine months putting them out.

Yet at their peak, the Kuwaiti oil fires consumed only 2 percent of the fossil fuels that humans burn every day, all day, year in and out. Imagine fifty sets of Kuwait-sized fires burning around the clock, never stopping, and you get a sense of what humanity is doing to power our industrial civilization.

Every person living in a well-off country contributes to the conflagration. When you and your neighbors turn on your lights at night, a coal- or gas-burning power plant somewhere will likely increase its fuel use—just a smidgen—to supply the electricity. Take a shower, and the natural gas in your water heater will fire up. Drive to work, and the engine in your car will burn the distilled remains of long-dead swamp algae just a few feet in front of your face at the rate of six thousand tiny explosions a minute—nicely muffled, mind you, but there all the same.

The clothes you buy, the warmth you enjoy indoors in the winter, and the coolness in summer—all these comforts are derived from the flames we hide away in chemical factories, power plants, furnaces, and engines. If energy consumption were measured in matchsticks, each American would strike nearly 5 million matches a week. Even in a country like China, much poorer but catching up, the figure would approach 2 million.

Without meaning to, we have been heating up the world, as the gases from all these flames alter our atmosphere, trapping extra energy from the sun. And while many of us find it hard to connect our own activities to this slow-motion emergency, we are starting to feel the consequences in our daily lives: heat waves worse than any in recorded history, rising seas flooding major cities, a runaway increase in wildfires that are burning down homes, polluting the air, and cutting lives short. Polar ice caps are starting to melt, and the once-icy tundra is catching fire. We may be putting the world's food supply at risk.

Humanity faces a profound moral and practical dilemma: How do we sustain the economic progress that has delivered billions of us from poverty—indeed, how do we extend that progress to those still suffering—while quenching the fires that threaten our only home?

Many people are already trying to help, in their own ways—perhaps by buying a Prius or an electric car, recycling diligently, installing smart thermostats, eating less meat, maybe contributing money to an environmental group. These actions are important, but by themselves they are not enough. The world will not be saved by conscientious *green consumers* who decide, one family at a time, to drive less or install solar panels on the roof. The problem is just too big for that.

Instead, we all need to become *green citizens*. We need to focus, together, on a relatively small number of public policies that can, over time, bring about sweeping change. And that requires a coherent plan, one that every concerned citizen, business leader, technical innovator, and politician can understand. With strategic clarity comes power—and, we hope, the will to accelerate the necessary changes and the skill to minimize the associated cost and risk.

The good news is that change has already begun to happen, in pockets across the globe. In Britain, for the first time since the nineteenth century, many weeks go by now without a single lump of coal being burned to generate power. The country that led the world into the Industrial Revolution—and its insatiable hunger for fossil fuels—now seems increasingly determined to lead us out of that dependency.

Sometimes the changes we need are hidden away in closets: In Oregon, thousands of water heaters are getting new digital controls to allow them to compensate for variability in the electrical grid. When electrical demand is high, they hold back their energy appetites; when it is low, they power up and then store that heat until it's required. This kind of innovation will help match demand on the grid to more variable supplies like wind turbines and solar panels. And nobody has to step into a cold shower.

Sometimes the changes are displayed across hillsides: in North Carolina, solar power has boomed, thanks to decisions by the state government to allow low-cost renewables to compete fairly. Soon, the tides of change will be visible in the sea: in the American Northeast, a huge new industry is gearing up to build wind turbines miles from shore. Changes are appearing in city centers: in Germany, swanky new apartments supply all the climate control their occupants desire using less electricity than a hair dryer. In China, a massive market shift toward electric cars and buses is underway.

In other words, solutions are already at hand, and more are coming. If they can be harnessed at a sufficient scale, neither our living standards nor our economy need be at risk as we transition to clean energy. But we are not moving fast enough to adopt them. An ominous combination of ignorance, inertia, and political mischief is hampering the pace of our energy transition.

At this moment in history, speed is the critical issue. The world's nations agreed, in a meeting near Paris in 2015, to try to keep the temperature of the planet from rising to catastrophic levels. They set a specific upper limit: a warming of 2 degrees Celsius, or 3.6 degrees Fahrenheit, above the average temperature before the Industrial Revolution. They also set an aspirational goal: to hold the increase to 1.5 degrees Celsius. Those numbers may sound low compared to the swings in temperature we experience day to day, but averaged across an entire planet, they are actually quite large. Unfortunately, the fossil-fuel combustion of centuries past has already heated the Earth by more than 1 degree Celsius, meaning we are already more than halfway to the danger zone. And we are not remotely on track to meet either of the Paris goals. If we blow past them, calamity awaits us on the other side. We are getting a foretaste of what it will be like in the fires already destroying towns in the American West, in the droughts that are drying up our water supplies, in the heat waves that are sending temperatures soaring past 120 degrees Fahrenheit.

Meeting the Paris goals means that the remaining amount of fossil fuel we can burn is limited. To meet the 2C goal, as it is known in shorthand, the era of fossil fuels must come to an end by around 2050—less than thirty years from now. To get on a path toward that target, we need to cut our emissions substantially in just the next decade. But at a global scale, emissions are not falling. They are still rising.

No country is on track to meet these targets, though many—especially in Europe—are trying. President Joseph R. Biden has also set ambitious new goals for the United States, but no matter how much he achieves, the climate crisis will extend far beyond one or two presidential terms. Like other countries, the United States needs to embark on a decades-long course of cutting its emissions. And no matter how much help comes from Washington, much of the work will need to be done by state governments and local communities.

You may have heard the argument that it is already too late to stop the coming disasters. In a sense that's true: the limit of two degrees Celsius was not picked because it is safe, but because it is doable—if barely. Given the fires, extreme rains, and coastal floods that we are already seeing, a world that has warmed up twice as much as ours is going to be a challenging place to live. Yet in another sense, it is never too late to act: so long as a pound of coal or a barrel of oil remains in the ground, we have agency. We can choose not to burn it, and our efforts will leave a better world. And so, the way we see it, it is not too late to tackle the climate crisis. We still have time to head off the worst damages, to prevent a great deal of human suffering, and to do right by future generations.

We're heartened by how many people are already doing what they can to help. Yet it can be difficult to know *what* to do: the problem is so big, and we all feel so small. And so our goal in *The Big Fix* is to lay out the grassroots political actions you can take that will have the greatest impact, because you do have the ability to make a difference. The problem certainly is enormous, but that just means that every single one of us can tackle a piece of it.

We may be optimists, but we're not naïve. Our backgrounds have given us a deep appreciation for the technological, economic, and political complexities of our world. Hal trained as a mechanical engineer—thirty years ago he built his own electric car, before you could buy them, and charged it with solar panels—and he has spent decades advising political leaders around the world on how to speed up the transition to clean energy. He knows from experience which policies work and which don't, and he knows how best to advocate for them. Justin, meanwhile, has had a forty-year career in journalism, including nearly a decade as the lead reporter on climate science for the *New York Times*. So he understands the power of a well-told story to inspire change.

In this book, we will not dwell on changes we regard as politically impossible. Many economists argue, for instance, that a hefty tax on emissions of greenhouse gases would go a long way toward solving this problem—and it might, if there were any chance of getting it done. But thirty years of efforts to pass modest emissions taxes (far too modest to do the job, really) have yielded nothing in Washington. So we're not going to suggest you spend your time and money trying to achieve something that history suggests cannot be achieved in the years left to us before we reach 2C. Instead, we'll focus on the actions that promise to give us all the biggest return for our investment of time and energy.

We've structured the book by breaking our economy into seven realms: six are the economic sectors that contribute most substantially to the emissions problem today, and the seventh is the realm of invention, both technical and financial, that can help cut future emissions. To save our climate, we'll need to make practical advances in each of these seven realms until we've brought our emissions down to near zero, and in each chapter we'll show you how you can exercise influence to steer things in the right direction. Cleaning up the electric grid is the critical first step, because clean electricity can be used to displace dirty fossil fuels in other parts of the economy. We'll dive into how our society can stop wasting so much energy in buildings and how to cut emissions in our transportation systems. The way we produce food and manage our land needs to change, too, and as people rush from the countryside to urban areas, we'll need to build cities that are more sustainable. We will cover ways that we can begin cutting the greenhouse gases spewing from factories that produce the goods we all buy. And while many of the changes we need to make are already clear, others are barely visible on the horizon—so in our last chapters, we'll show you what society needs to do to speed them up. In our book title, we call these "steps," but we do not mean to imply that they need to happen in a particular order. Society needs to pursue all of them at once.

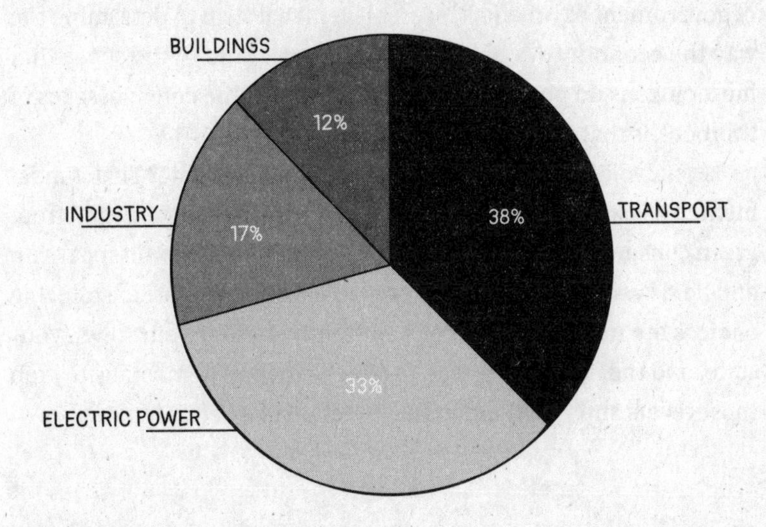

Technology will play a large role in the fight to save our climate. But what we really need, even more than new technologies, are better policies to ensure we implement these technologies widely enough to make a difference. You might think that such decisions are made by businesses and governments, with little opportunity for citizen input. But that's not quite true. When your town decides how strong the local building code will be—and so how much energy new buildings will be allowed to waste, for decades to come—you can bet many local builders ply their influence to try to get the weakest code they can. You, as a citizen, can ply your influence, too. The board in your state that makes the final decisions about what kind of power plants get built is likewise required by law to listen to the public and take its interests into account. And your state's elected officials have more influence than you might realize about everything from what kind of cars will be sold on the local market to how efficient appliances have to be. It's dozens of actions like these, carefully chosen and then intensely pursued, that are, collectively, the key to solving the climate crisis. You can think of these decisions being made at various levels of government as, in effect, secret levers that help to determine the way the economy develops. They are secret only in the sense that most citizens do not know about them; one of the core messages of this book is that the time has come to learn.

Is your voice being heard as these decisions are made? That, fundamentally, is what this book is about: how to make the transition from green consumer to green citizen, becoming someone who speaks up and holds your government accountable, ensuring more sustainable choices are made not only for your family but for your town, your state, and the nation. We need to use the tools of democracy to grab those levers and pull them in the direction of a better future.

CHAPTER 1

The Learning Curve

A sturdy workboat called the *Alliance* cut swiftly through waters a few miles off the English coast, riding the calm seas with ease. In the stern of the ship stood a sprightly fellow named Julian Garnsey, a Briton who was narrating the journey to his passengers. As he spoke, he was peppered with questions by several men representing large Japanese utility companies. Big money, smelling opportunity, had come to see the world's latest technological boom.

Not so many years earlier, Mr. Garnsey, an engineer, had launched his career building oil platforms in the sea. But he let himself do some hard thinking about the implications of burning oil. Now he builds wind farms in the sea. His company, called RWE Renewables, is in the forefront of a dynamic new industry.

When offshore wind farms started to pop up in European waters in the 1990s, they were derided as a harebrained scheme with little chance of becoming a part of the energy mix. New, specialized boats were needed to pound piles into the seabed. Workers had not been trained in the complex techniques required to install giant machines on the ocean floor; investors were wary, and so demanded high interest rates to make construction loans. The cost of putting turbines into the sea was so crippling that companies demanded huge subsidies from European governments just to install a handful of projects.

Those governments were tempted, nevertheless, by a compelling physical reality. Wind turbines on land had long proven themselves to be a useful way to generate electricity, but the breezes over land could be fickle. Over the ocean, the wind blows harder and it blows more steadily, which meant that offshore turbines could produce more power—if they could withstand the harsh marine environment. The early, costly projects proved the theory, and as the industry scaled up, costs began to fall.

These days, the offshore wind industry has become one of the world's hotbeds of innovation and ambition. Americans have been touring Europe to see how the technology works; back home, legislatures and governors in the Northeastern United States are ordering up massive offshore wind projects. The Chinese are trying to seize a big share of the turbine market. The Danes, who more or less invented the industry, are fighting to hold on to their market share. The Germans, the French—everybody sees now that this is going to be a vital industry. The latest analyses show, in fact, that it may ultimately be capable of supplying a large fraction of the world's electrical power. It could turn out to be the most important industry created entirely in Europe since the end of World War II.

How did this happen? The short answer is those falling costs. When European governments first began to commit to offshore wind farms, they ordered their utilities to spend four or five times the prevailing electricity cost to acquire power from this new industry, with the extra cost passed along to electricity customers. But lately, contracts have been signed for wind farms to come online in the early 2020s at costs roughly in line with market prices, meaning that the offshore wind industry has learned to build its projects with little or no need for subsidies.

Offshore wind is just one example of the drastic reductions in the cost of clean-energy technologies happening all over the world.

Since 2010, the price of electricity from utility-scale solar farms has fallen almost 90 percent. Onshore wind fell 60 percent in the same time. Advanced batteries, which power electric cars and are increasingly finding a role in balancing fluctuations on the electric grid, fell more than 80 percent. When highly efficient light bulbs made with light-emitting diodes, or LEDs, first came out just over a decade ago, you could easily pay $50 for one; nowadays they sell at Home Depot for $1.24 apiece, a decline of 97 percent.

These technologies are all marching down a slope called a learning curve. As the market for them scales up, they keep getting cheaper. For most of the things you buy, like milk or a haircut, large price declines like these are not the norm. In fact, the day-to-day cost of living generally goes up, not down. Understanding the special economics that applies to certain energy technologies—both those mentioned above and, potentially, to new ones just being invented—is the key to saving the world from the ravages of climate change.

Every clean-energy technology, in its early stages, is costlier than the conventional alternatives. That makes the new ones a hard sell even in rich countries; all the more in the developing countries, like China and India and Indonesia, that will produce most of the world's future greenhouse emissions. You sometimes see politicians argue for laboratory research as the way out of our emissions dilemma—and we certainly do need more of that. But inventing new technologies is not enough. Any new technology has to become affordable if it is to be used widely. And so we need savvy tactics, including stronger public policies, that will drive these alternative technologies to scale—and make them cheaper in the process.

Any aspiring green citizen needs to understand the purpose of the public policies and private actions we advocate in this book. We are trying to make low-emissions technologies so cheap they become the default choice nearly all the time. The costs need to fall to the point where poor countries striving toward the global middle class can skip fossil fuels and go straight to clean energy.

That is our tactical goal: to make clean energy unbeatable in the marketplace, no matter how much political skullduggery the fossil companies might gin up. To understand exactly how to do that, we need to go back in time a bit—to figure out how certain technologies like wind and solar have become cheap over the past century. This historical tour may seem like a digression from the urgent task of advocating change today, but in truth, it's the template for what we must achieve. We need to take these lessons from the past and apply them to the future.

Julian Garnsey, the engineer from RWE Renewables, has a special way of showing how much things have changed in offshore wind even in just the past decade. As the *Alliance* plowed through the gray waters off the Essex Coast on a summer's day in 2019, moving into deeper seas, it entered the edge of a wind farm called Greater Gabbard. Greater Gabbard—built by a predecessor company of RWE and named, like many British wind farms, after a nearby sandbank—was one of the early offshore wind farms for which the British government had effectively guaranteed prices well above the market rate. The boat passed turbine after turbine, most of them motionless, although a few turned lazily as they caught the gentle summer breeze. People who see pictures of offshore wind turbines routinely misjudge their scale; it is easy to do with no trees or buildings on the horizon for contrast. In fact, the machines are as tall as skyscrapers. The Greater Gabbard turbines lined up in rows, like giant soldiers marching through the sea, stretching so far to the north and south that it was impossible to see all 140 of them from the boat at the same time. Yet they are but a fraction of the offshore turbines Britain has built, more than 2,000 of them—and the nation is just getting started.

Soon the vessel crossed an invisible line. It had left Greater Gabbard and entered a new wind park called Galloper, one that Mr. Garnsey and his team finished building in 2018. To the naked eye, nothing looked much different, but in fact the technology had changed markedly in a few years. These turbines were larger than the ones in the neighboring park just to the west. They were taller, the blades were longer, and they could capture more power from the wind—making each turbine capable of producing 75 percent more electricity than the older model. The newer of the two wind parks had only 56 turbines, which were faster and easier to install than the 140 older ones had been. And all of this meant cost savings that showed up in the price of electricity from the new park.

The boat slowed, and gradually approached one of the turbines that Mr. Garnsey's team had installed in the seabed. The great machine was mounted on a shaft that rose from the sea, the first few feet above the waterline painted yellow as a warning to boats. Up, up, up it soared, as tall as a sixty-story building, terminating in a structure at the top that was the size of a small house. Inside that structure, called a nacelle, the electrical generator was hooked up to a hub in the front. Attached to that hub, in turn, were three long blades capable of catching the wind and turning the generator shaft. The blades were hollow, but made of advanced materials, including fiberglass and carbon fiber, to give them strength enough to withstand a North Sea gale. Mr. Garnsey explained that cables snaking across the seabed were collecting power from the turbines and carrying it to shore. The bodies of the turbines can withstand harsh winds, too, in part because the shafts on which they are mounted are hammered deep into the seafloor. As the boat hovered below the turbine, Mr. Garnsey chatted with his Japanese visitors, potential investors in future projects of this sort.

"The amount of interest you get talking to someone about offshore wind farms is just incredible," he said a bit later. "They just get fascinated by the engineering. They start asking, 'Well, we're in the middle of the sea! How does this thing stand up? How did you drive it into the seabed? How big was the hammer?' " The answer to that last question tended to produce gasps: the hammer on this project was the size of a three-story building. You need a big boat to handle that hammer. One of the ways the industry has cut costs is by building its own specialized boats. The construction of a wind park can put a fleet of boats and a thousand people on the water at once. "If you have bad weather one day and nobody can work, that's a million pounds gone," Mr. Garnsey said; that sum is equivalent to $1.3 million.

Turbine H6F had caught the breeze and was turning lazily, operating at less than 10 percent of capacity. The low electrical output was not a problem, though: the pleasant summer climate in Britain requires little air-conditioning, and demand on the national grid was minimal. The output from these turbines would be needed most in the winter, when Britons strain the grid to heat their houses. Fortunately, the North Sea winds blow hardest in winter. As he stood under the turbine, Mr. Garnsey pointed out that a single rotation of the blades would produce enough power to run an electric car for thirty miles. In a year, that single turbine would supply enough power for more than six thousand British homes, keeping the lights on, the washing machines running, and the tea kettles humming. The British prime minister, Boris Johnson, recently vowed to keep building wind parks until every household in Britain can be supplied with clean electricity from the sea.

As the boat turned back to shore, Mr. Garnsey began to speak of his next project, Triton Knoll. The turbine size will jump another 50 percent, and the power will again get cheaper. The manufacturers of turbines—companies like Vestas in Denmark and General Electric in the United States—are racing to see how much bigger they can make these machines. Vestas has announced it will build a turbine capable of generating fifteen megawatts of power, twice the size of the massive turbines that were being installed only recently. The blades of the machine will trace a circle so immense that two Airbus A-380s, the largest passenger plane in the sky, could fly through the circle side-by-side—with room left between them for a half dozen American fighter jets.

While Britain has built nearly a third of the world's offshore turbines, other European countries bordering the sea have also played major roles. The United States has not. It watched as this new industry developed abroad. But the plunging costs have finally awakened American interest. The United States has only seven turbines operating off its coasts now, five in Rhode Island and two at a test site in Virginia. But many new projects have lately gone to bid in the United States.

The earliest attempt at developing an offshore farm in the United States, in Massachusetts, was proposed more than twenty years ago. It was too close to shore and was thwarted by opposition from nearby property owners, but the technology has since developed to the point that turbines can be installed beyond the horizon, making them invisible from people's beach houses. This technological development has also been a crucial political development. State governments, with federal help, are now aiming to put thousands of wind turbines in the shallow continental shelf off the coast of the Northeastern United States. These states have already planned enough offshore wind to generate as much electricity as five or six nuclear power stations would produce, and the Biden administration has called for multiplying that threefold by 2030. A national commitment to the technology is critical, because it is the federal government that controls the seabed beyond three miles from the coast, and so only it can grant leases to wind-farm operators.

The falling cost rippling through the offshore wind industry—and onshore wind, and solar power, and LED bulbs, and electric cars—may seem like some kind of magic trick. But in reality, certain economic rules are at work, and they are reasonably well understood by specialists. The cost declines in these industries were entirely predictable; indeed, they were predicted, in some cases, decades ago. To understand what is going on with these technologies—and, by extension, what society needs to do to develop and adopt those of the future—we need to go back in time by a century, to the early days of one of the modern world's defining machines: the airplane.

Wright's Law

Theodore P. Wright had a head for numbers. He had trained as an architectural engineer, but as a young man, he got pulled into the navy just as the United States was entering World War I. He was immediately put to work on problems involving primitive war planes, then a naval responsibility. Within months he was publishing technical papers and navy manuals stuffed with equations, solving arcane problems of aircraft design and construction while still in his early twenties. He had found his calling, and a newborn American industry had found one of its native geniuses.

At a time when trains dominated long-distance travel, Mr. Wright was one of the first people to envision a passenger airline industry operating on a large scale. He would go on to become a critical figure in helping the United States produce enough airplanes to win World War II, and after the war, he helped bring his early vision to life, serving as the director of the agency that would eventually become the Federal Aviation Administration.

For all those accomplishments, Ted Wright might well be a footnote in history, save for one brilliant insight. Early in his career, he was trying to understand changes in the cost of building airplanes. It was obvious that the costs fell as production scaled up and more and more units were produced, but was there any pattern that made that cost decline understandable or, better yet, predictable? He later wrote that he had found a pattern by the early 1920s, but for a time, the company he worked for, the Curtiss Aeroplane and Motor Company, kept his discovery as a trade secret. All through the 1920s, the company used Mr. Wright's method as it prepared bids to sell airplanes to the government and other buyers.

"Then, in 1936, I went to Germany and had my eyes opened; saw production of military aircraft in numbers many times the combined output of the United States and the United Kingdom, and very obviously, not intended for national defense, but for aggressive war," he wrote years later. "I was regretfully forced to accept the fact that the democracies, having already lost the peace that followed the First World War, must, if they were to survive, prepare intensively for the second. Our potential enemies had selected the weapon; their strength was great and time was short."

Just before he took that trip, perhaps already thinking hard about what the United States would need to do if war came, he decided to spill his company's secret. The paper he finally published, "Factors Affecting the Cost of Airplanes," ran in February of 1936 in the journal of a trade organization, the Institute for Aeronautical Sciences. Surely no more than a few hundred people read it at the time. But in the decades since, it has come to be viewed as one of the more important papers in the history of manufacturing.

By the early decades of the twentieth century, industrialists already understood that as they scaled up to make a new product, the cost of building it was likely to fall. The most famous example was Henry Ford and his Model T automobile, the first mass-market car in the United States. When the Ford Motor Co. first brought the car to market in 1909, it built just under eleven thousand of them, and the most popular variant cost $850 apiece. By the peak of the Model T in the mid-1920s, Ford was building nearly 2 million cars a year, and you could buy one for less than $300. As the Industrial Age brought more and more consumer goods to the public, that pattern—an initial scale-up accompanied by a sharp decline in the cost of each unit produced—was seen repeatedly.

A fundamental reason was the principle known as economies of scale, an idea rooted in eighteenth-century economics. Mr. Ford needed a factory and equipment and a minimum number of workers, no matter how many cars he was going to produce in a year. He would incur costs for design and testing and general company administration. As the business scaled up, he could spread these fixed costs over more and more cars; the cost of producing each car would thus be lower. Moreover, as production grew, his workers became faster and more skillful at their jobs. The company kept developing new tools and machinery to cut time and costs. Its most famous innovation was the moving assembly line, an idea Mr. Ford and his team got when they visited the Chicago stockyards. They had watched the way rows of laborers disassembled a carcass rolling by on a hook, each worker slicing off the same part over and over. In essence, Ford and his team turned that *disassembly* line into an *assembly* line, using a conveyor belt that enabled each worker to attach the same car part over and over. Monotonous it may have been, but it was fast and efficient, cutting the labor cost of making automobiles even as Mr. Ford offered his workers excellent pay for the era. The improvements at the Ford Motor Company showed up in the lower price of the cars, propelling the Model T to the top of the public's wish list and launching the age of the automobile.

Here is a simpler way to think about economies of scale: Have you by chance ever tried brewing beer in your basement? It's a fun hobby, but you probably figured out pretty quickly that you would not be saving any money on beer. You may have spent hundreds of dollars buying a beer capper, a fermenter, tubes, and so on. You might have spread that steep investment cost over, say, a case of beer a month. Most of the time, your pricey new gear probably sat around unused: a wasting asset, in economic terms. Somebody with $200 million invested in a brewery, running day and night to make 100 million bottles a year, is always going to beat you on the cost of producing beer. However delicious your lager might be, economies of scale give that factory owner the advantage on cost.

By the early twentieth century, the idea of economies of scale had become a part of basic economic theory. Ted Wright's father was an economist, so Ted would certainly have understood it when he tackled the problem of airplane costs. But what he wanted to know was how *much* costs would decline for a given volume of production. This was a subtle problem, for the answer would depend not just on scale economies but on how much better workers became at their tasks as production ramped up, on engineering tweaks that might make an airplane easier to produce, on lower prices for supplies as the volume scaled up, and potentially a hundred other factors. Yet finding an answer might offer Mr. Wright's employer, Curtiss Aeroplane, a big advantage in bidding on government orders: if it could predict the fall in costs accurately, it might underbid its competitors.

Mr. Wright finally found his answer on a sheet of graph paper. When he plotted past production increases against the amount of labor it took to build airplanes, he found that every time production doubled, the labor requirement fell by about 20 percent. Workers were learning how to do their jobs more quickly and efficiently, and management kept rejiggering the layout of the factory floor, shortening the distance parts had to travel and improving overall procedures. After three doublings of output, labor costs had fallen by almost half. The amazing thing was the regularity of it. This idea—that costs of factory production will fall by a more or less fixed rate each time total historical production doubles—is now known as Wright's Law. It has been found to apply not just to labor, but to many aspects of factory production. Here, again, is the classic example: the cost to the public to buy Henry Ford's Model T automobile.

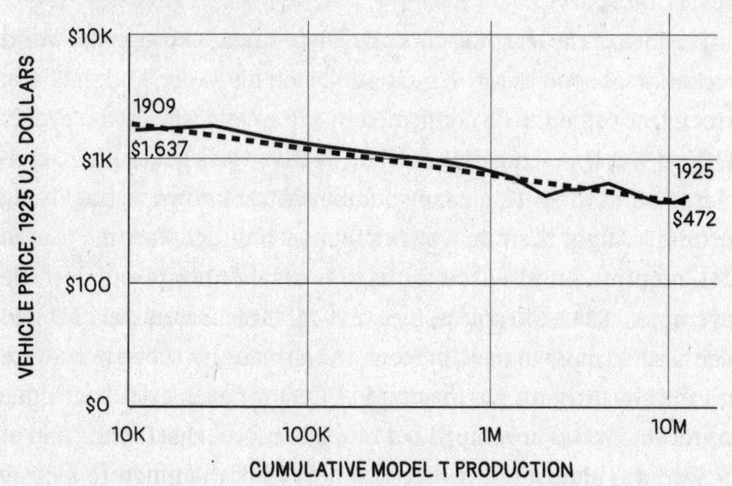

UNLIKE THE FIGURES CITED IN THE TEXT ON PAGE 17, THESE ARE ADJUSTED FOR INFLATION AND INCLUDE ALL TRIM LINES OF THE MODEL T.

Note the numbers across the bottom of the chart showing the rapid increases in production as public demand for the car soared. This chart demonstrates that the Model T displayed a "learning rate" of around 15 percent, shown by the dashed line. That is, every time cumulative production doubled, the costs fell by that much.

Most people have never heard of Wright's Law, but the idea eventually permeated the business world of the mid-twentieth century. The sloping line you get when you make one of Mr. Wright's charts is called a learning curve, or sometimes an experience curve—various terms have been used over the years, sometimes with varying definitions, but the basic idea is the same.

In casual speech, people use "learning curve" to mean something that is difficult. One might say, "I studied Russian in college, but it was such a steep learning curve, I never mastered it." The learning curve associated with Wright's Law is distinctly different. The line on his graph works more like a child's slide at a playground: you get on at the top and go down. An article of manufacture that is getting cheaper as production scales up is said to be moving "down the learning curve."

The idea of the learning curve began to spread through the world of aviation as soon as Mr. Wright published his paper, and the basic principles seemed to be confirmed in a dramatic way by the events of World War II. A struggling outfit in Seattle won a government contract to build the B-17, a heavy bomber better known as the Flying Fortress, to fight the war. The first planes built for wartime use, in 1941, required hundreds of thousands of labor hours and cost the government $242,200 apiece. By late 1944, the labor requirement had been slashed more than 90 percent, the airplane had been improved, and the government was paying $139,254 for each one. That huge improvement was accomplished by a labor force that, by the end of the war, was almost half-women, almost all of them new to factory work. This was not the only example of massive cost declines during the war, but it became one of the most famous. The Flying Fortress dropped more bombs on Nazi Germany than any other plane. Wartime manufacturing prowess turned the Seattle company that made the plane into a household name in the United States: Boeing.

To this day, scholars still study the manufacturing experience of World War II, because it is just about the only way to answer a critical question regarding technological learning rates: Are items really getting cheaper *because* manufacturing scales up, or is manufacturing scaling up because prices are getting cheaper, leading people to buy more and more of an item? In ordinary times it is nearly impossible to disentangle cause and effect, but World War II clarifies the direction of causality. "We know that Franklin Roosevelt was not buying more airplanes and tanks because they were getting cheaper," said J. Doyne Farmer, a researcher at the University of Oxford who works on technological learning rates. In other words, an extreme threat to Western civilization—the rise of the Nazi tyranny in Europe—led to a sudden, drastic increase in demand for the matériel of war. As American factories retooled to meet that demand, their costs fell in approximately the way that Mr. Wright's learning curve predicted they would.

The implication, for modern times, is that any deliberate effort to scale up a new technology might reasonably be expected to lead to falling costs. And remember that what matters for the learning rate is *cumulative* production, not time. The faster production doubles and doubles again, the faster the costs ought to fall.

After the war, the newly created U.S. Air Force adopted the idea of the learning curve as a fundamental way to understand how production costs might shift over time. It found heavy use during the Cold War as a tool of military procurement. A related idea took hold in the computer industry. Gordon Moore, a cofounder of the Intel Corporation, observed that the number of transistors that could be squeezed onto a computer chip appeared to be doubling every two years, which meant the effective cost of computing power was being cut in half every two years. Researchers have established that Moore's

Law and Wright's Law are closely related; computer chips are a technology with a learning rate similar to that of many other technologies. But they have proven to be so useful that the production has doubled many, many times, so that the learning rate has led to precipitous price declines.

In the middle decades of the twentieth century, a business school student could not graduate without some exposure to the idea of the learning curve. As a description of what happens in entire industries, it made plenty of sense. But proponents of the idea went overboard when they tried to use it to guide the strategy of individual companies. What people most wanted to do with learning curves was to predict future costs, but it turned out the slope of the line *could* sometimes shift. In other words, your nice, regular learning curve could drop from 20 percent to 10, or perhaps it might jump from 20 to 40. Worse, running a manufacturing business with a healthy learning rate turned out to be no guarantee that some competitor would not disrupt your business model with an improved product or an entirely new invention. Simply put, the analysis of technological learning rates was no surefire path to success in manufacturing. Moreover, in the service industries that increasingly dominated the economy in the second half of the twentieth century, the idea was of little use. Law firms did not seem to be willing to charge less for pumping out real-estate contracts no matter how many times they did it. By the late 1980s, these seeming imperfections of Wright's Law had pushed it aside as a workaday tool of business.

But an urgent new problem was moving onto the world's agenda, and this old idea was about to be put to a new use. To understand the story of clean energy in the late twentieth century, it helps to grasp a corollary of Wright's Law. If pushing technologies to scale could bring the costs down, then the obverse was true, too: technologies left to languish would never move down the learning curve, meaning they might never become cheap enough for widespread use. Two technologies serve as prime examples: wind turbines and solar panels, both used to generate electricity. For as much excitement as they are causing today, each of those technologies languished for many decades of the twentieth century. That was a fateful mistake, and understanding this is critical to tackling the climate crisis, for it is a mistake society cannot afford to repeat.

Governor Moonbeam

Had you met Dew Oliver in 1926, you might have written him a check. A lot of people did, and came to regret it. He was a charming Texan running around Southern California in a cream-colored Stetson cowboy hat, sporting a walrus mustache and talking up moneymaking schemes. His boldest idea was a plan to capture the wind.

Mr. Oliver, like just about everybody else who passed through the San Gorgonio Pass, was mightily impressed by the winds there. The pass, created by the famed San Andreas Fault, is one of the steepest in the United States, with the mountains on either side rising nearly nine thousand feet above it. Like a lot of mountain passes, it functions as a wind tunnel. As the hot desert air of interior California rises, cooler air from the Pacific Ocean, to the west, rushes through the pass. The story goes that Mr. Oliver realized how strong those winds were when they blew his Stetson off his head.

His scheme was pretty simple, really. He wanted to erect a ten-ton steel funnel to capture the wind, then send it through propellers connected to a 25,000-watt generator. His intent was to sell the electrical power to the nearby budding resort town of Palm Springs. He apparently failed to realize that a local utility had already claimed the town and would not welcome an interloper. But he did get the thing built: by 1927, Mr. Oliver's wind machine had been erected at a spot a few yards from where Interstate 10 passes today. A huge funnel on the front end was attached to a cylinder seventy-five feet long and twelve feet wide, with propellers inside to drive a secondhand generator Mr. Oliver had scrounged up. But even Mr. Oliver had underestimated the power of the wind: in the early testing, a propeller spun too fast and set the first generator afire. He found a bigger one. Yet the few customers he managed to sign up complained that the power from his machine was erratic. Needing more money to improve his equipment, Mr. Oliver undertook to sell stock to local people, and it seems he may not have been entirely honest with them about the risks of his venture. One suspects the costs got away from him, but whatever the cause, the scheme failed. Mr. Oliver was hauled into court and convicted of selling stocks unlawfully. After a short stint in jail, he fled California, and his machine stood forlorn in the desert for years, eventually to be cut apart for scrap in World War II.

Why would any investor be duped into writing checks for such a crazy plan? Actually, the notion of generating electricity from the wind was a hot idea in the 1920s, and many Americans had read about it, if not seen it working. On thousands of farmsteads that had not yet been connected to the electrical grid, families were eager to gain access to the new medium of the age: radio. This new technology had soared in popularity in the mid-1920s, with five hundred new broadcasters going on the air in a single year, 1923. In the pre-radio era, farmers had gotten along with kerosene lanterns at night and no electrical power, but many now felt they had to get connected to the modern world. For one thing, critical farm news, including daily prices, was now being broadcast on the radio. Startup companies plied the countryside, selling kits that included a small wind turbine connected to a generator, a set of batteries, a radio, and an electric light or two. The devices were called wind chargers, and they were finally rendered obsolete in the 1940s, when one of Franklin D. Roosevelt's New Deal programs delivered nearly universal access to the power grid. Many decades later, though, the cultural memory of the wind chargers would prove to be important. Deeply conservative people living in the middle of the country, who might have been expected to oppose such newfangled inventions as large commercial wind turbines, remembered hearing about wind chargers from their grandparents. The idea of harvesting the wind, the way you harvested a crop, would strike many of them as a perfectly sensible thing to do.

By the time the wind-charger business collapsed in mid-century, it was clear you could generate significant amounts of electrical power from the wind. A few people had the vision to see how much bigger wind power could become: with extensive support from the Massachusetts Institute of Technology, a large-scale turbine was built in this era to feed electricity into the power grid. The turbine, installed atop a Vermont mountain called Grandpa's Knob, operated intermittently but successfully for five years, sending power to the Champlain Valley below. The turbine broke near the end of World War II, and since power from the wind was somewhat more costly than power from conventional generators, the local utility decided not to pay for new turbines. Yet a dream had come to life, and it would not die. The most important scientist in American public life of that era, Vannevar Bush—who had been President Franklin Roosevelt's science advisor during World War II—had kept a close eye on the project.

"The great wind-turbine on a Vermont mountain proved that men could build a practical machine which would synchronously generate electricity in large quantities by means of wind-power," Dr. Bush wrote in 1946. "It proved also that the cost of electricity so produced is close to that of the more economical conventional means. And hence it proved that at some future time homes may be illuminated and factories may be powered by this new means."

While Dew Oliver's project to generate wind power in the desert had come to naught, he had gotten one thing right: he had indeed found one of the best places in the nation to capture the wind. Half a century after his scheme went under, the idea of generating power at commercial scale with wind turbines would be reborn, and the San Gorgonio Pass would be one of the places where it happened. California has two other mountain passes, the Tehachapi in the south-central part of the state, and the Altamont in the north, with similarly windy conditions, and they would also prove important.

As we will discuss in more detail later in the book, the energy crises of the 1970s came as a huge shock to the American people. A country where energy had seemed boundless was suddenly afraid of running out. By the late 1970s, California had a young governor, Jerry Brown, and the United States had a president, Jimmy Carter, who were willing to look beyond fossil fuels for new energy sources. Global warming had yet to become a major issue, but the pollution from fuel-burning had become a big one, and so had the fear of running out of oil.

Mr. Brown, elected governor at the age of thirty-six, was an especially creative thinker. He had spent some of his younger years in a Jesuit seminary, enmeshed in the world of ideas. Soon after Mr. Brown took office in 1975, his imagination was captured by a fat tome published the year before: *Energy for Survival: The Alternative to Extinction*. The author, Wilson Clark, warned that human society was on an unsustainable path with its rising demand for fossil energy. Aside from pollution and other near-term problems, he warned of "a warming of the Earth that would cause melting of the ice caps and associated natural disasters across the globe."

Jerry Brown did more than admire Mr. Clark's book: he drafted the man as his new advisor on energy. The governor's office became a hothouse of discussion about the future and how California could help to invent it. Looking askance at some of the ideas emanating from the state capitol at Sacramento, a Chicago newspaper columnist named Mike Royko dubbed Mr. Brown "Governor Moonbeam"; the nickname stuck. In 1976, the audacious Mr. Brown ran for president at age thirty-eight, losing the Democratic nomination to Jimmy Carter. They remained political rivals through Mr. Carter's four years in office, but they were aligned in many ways on energy policy. In the search for energy alternatives, both the state of California and the federal government opened the spigots of government finance. Mr. Carter also signed a federal law requiring that utilities buy power from small generators, including wind farms, if they could beat the generating costs of the big companies. That created a nascent market for renewable energy.

Many of the energy ideas from that era would end badly, including a massive effort to turn coal into liquid fuels, but some of them would come to fruition. Commercial wind power was the most important. The government subsidies and guarantee of a market prompted a mad rush to put up experimental wind turbines in the three mountain passes of California. Nobody repeated Dew Oliver's idea of a huge funnel; engineers understood by then that the best way to capture the wind was just to stick up blades attached to a generator and let them turn. They did try all sorts of designs, some of them, in retrospect, nearly as crazy as Mr. Oliver's. Turbines that looked like huge upright eggbeaters were tested, but had a tendency to rip themselves apart. Engineers tried using two blades, and five blades, in multiple sizes perched atop various kinds of towers. More than one dreamer and schemer jumped into the industry, ultimately failing despite the government largesse.

On the far side of the Atlantic Ocean, the same search for alternative energy sources was going on, and for similar reasons: European countries, some of them completely dependent on fuel imports, were shocked by the vulnerabilities that the Arab oil embargo had laid bare. The French locked in on nuclear power as the solution, building dozens of reactors and burying the costs in the state budget. In Denmark, a flat country surrounded by water and whipped by stiff ocean breezes, many people zeroed in on the wind. It would be the Danes who finally built what is often regarded as the first modern wind turbine.

It was not some Danish university or government research program that solved the puzzle, however. What happened was far stranger: a controversial organization called Tvind, a conglomeration of alternative schools and businesses that has sometimes been labeled a cult, designed and built that first big turbine, motivated by an ideological opposition to nuclear power and fossil fuels. With volunteer labor, the group erected a machine as high as a fifteen-story building, with the blades sweeping an area nearly as wide as the machine was tall.

The turbine, which went into operation in 1978, had three long blades, the form factor that would ultimately become the global standard. That design had already been shown to work on the American Great Plains, but the Danes scaled the machine up immensely. The project received extensive news coverage, and tens of thousands of Danes flocked to see it out of sheer fascination. So began Denmark's political commitment to developing wind power.

Ideas flew back and forth between California and Denmark, with the wind developers of California ultimately adopting the three-bladed Danish design, which seemed to offer the best balance between aesthetics and engineering. For a while, Denmark was exporting more turbines to California than it was installing at home. Thousands of turbines had gone up in the California mountain passes by the time Mr. Brown left office and the spigot of government support was turned off. By the standards of later decades these turbines were still primitive, but they proved that a battery of turbines—a wind farm—could supply electrical power in serious quantity.

In retrospect, by the early 1980s, the United States was on its way to developing a major new source of electrical power. But then Ronald Reagan, taking office in 1981, largely ended the national commitment to clean energy, embracing expanded drilling and mining for fossil fuels. A stalwart handful of wind-power companies survived the shakeout and worked to perfect their designs, but in the Reagan era, with the oil crisis over, most Americans were paying little attention to energy security.

Around this time, another technology was starting to take off, too. And its history demonstrates that wind power was not an anomaly, either in its ability to generate large amounts of power or to undergo huge cost reductions.

Just like wind turbines, solar panels had a long prehistory. It had been clear since the nineteenth century that if you exposed certain materials to sunlight, you could generate electrical power. In 1905, Albert Einstein explained what was happening with this "photoelectric effect," work that would win him a Nobel Prize. Thomas Edison, the famed inventor, rhapsodized in 1910 about capturing the sun as a useful source of electricity. "The scheme of combustion in order to get power makes me sick to think of—it is so wasteful," Mr. Edison said. Throughout the first half of the twentieth century, scientists tinkered, trying to make a useful solar panel. In 1931, *Popular Science Monthly* interviewed a German researcher, Bruno Lange, who was working on solar cells. The magazine wrote that "in the not distant future, huge plants will employ thousands of these magic plates to transform sunlight into electric power . . . that can compete with hydroelectric and steam-driven generators in running factories and lighting homes." Prescient that claim may have been, but for decades, nobody could make a device that was efficient enough at converting the sunlight that hit it into electricity. That meant any attempt to use solar panels on a large scale would have been prohibitively expensive.

The breakthrough came in 1954 at a venerable American research institution known as Bell Telephone Laboratories, in New Jersey. American Telephone and Telegraph, the national telephone monopoly known as Ma Bell, was then one of the richest corporations in the country, and it funded a lavish laboratory that made some of the seminal technological breakthroughs of the twentieth century. An inventor at Bell Labs named Daryl M. Chapin was working, with two colleagues, on ways to supply backup power to remote telephone switches, potentially for use in areas of Latin America that the power grid did not yet reach. The group turned to solar cells and discovered that a plate of silicon—derived from sand—could, if mixed with tiny amounts of other materials, produce power from sunlight six times as efficiently as any previous cell. Bell Labs held a news conference at which the inventors powered a small Ferris wheel with their new solar panels. A writer for the *New York Times* declared on the front page of the next morning's paper that the invention "may mark the beginning of a new era, leading eventually to the realization of one of mankind's most cherished dreams—the harnessing of the almost limitless energy of the sun for the uses of civilization."

Just as with wind turbines, the development of solar panels would not happen quickly, and the problem in both cases was the same: nobody had a motivation to take on the costly task of pushing these technologies to scale. Compared to digging up black rocks and burning them, the devices were still an extremely expensive way to produce electricity. In essence, they were stuck at the top of the learning curve, not yet moving down it.

Technologies stuck in that position can sometimes survive by finding niche markets. That is what happened with solar cells: NASA realized early on that they might be an ideal power source for satellites circling the Earth, and in the space program of the 1960s, cost was no object. Gradually, the market expanded as other uses were found. Among the corporations that adopted solar cells early were oil companies, who discovered that they could be coupled with batteries to power navigation lights on offshore oil rigs. In fact, oil companies were some of the earliest big investors in the solar industry. By the 1980s, Japanese calculator companies had taken up the baton, and for years they were the principal commercial makers of solar cells. Seeing potential in this growing industry, utilities and the Japanese government launched a program to get people to put solar panels on their roofs. Late in that decade, a few intrepid Americans began spending large sums to put solar panels on their houses—often, homes that were too remote to connect to the electrical grid. Gradually, from one niche market to the next, solar power began to scale up. Commercial wind turbines also grew slowly in this era, as interest spread beyond Denmark and California to a handful of other countries, including the industrial powerhouse of Germany.

In 1988, a NASA scientist, James E. Hansen, warned Congress about the risks of global warming, drawing extensive news coverage. Over the coming decade, environmental groups began looking for ways to cut high emissions from the electric grid. In theory, nuclear power would have been one way to do it, but costs were escalating and public concern about the safety of nuclear plants was high at the time. What other technologies were available? Wind and solar power were obvious candidates, but their costs were still high.

However, as early as the 1970s, a handful of researchers had noticed that solar panels seemed to be getting cheaper as the market for them scaled up. The charts implied that if environmental advocates could figure out a way to make the market larger, that would drive the costs down further—but nobody wanted to sign up for these new technologies until the cost came down. Mr. Wright had understood this kind of paradox, writing in his famous paper in 1936: "We have the usual circle of relationships wherein price can be reduced most effectively by increasing quantity but wherein quantity production can only be obtained through market possibilities brought about by cheaper prices."

Not until 1993 did somebody finally make the next leap of logic. A professor at Princeton University, Robert Williams, pointed to the robust 20 percent learning rate for solar panels and suggested a strategy. If public money were used to scale the technology up more rapidly, then the costs should fall more rapidly, too. He proposed a big federal program to achieve this, but as we shall see in the next chapter, what actually happened would turn out to be messier and more complicated than that. It was, nonetheless, a brilliant flash of insight: the learning curve had graduated from being an observation about the way technologies behaved to the centerpiece of an effort to solve one of the world's greatest problems.

The Big Ask

It was the hottest ticket in London that spring. Only a thousand people were able to squeeze into KOKO, the famous venue in Camden Town that has hosted intimate concerts with musicians like Prince, Lady Gaga, and Oasis. The headliner for the May 1, 2006, gig was Thom

Yorke, the superstar lead singer of the band Radiohead, playing with his bandmate Jonny Greenwood.

This was no ordinary concert, though. Its purpose was political, and the organizers were careful to spread some of the tickets around where they might do the most good. Standing in the crowd at KOKO that night, soaking up the vibe, was a rising star of British politics, David Cameron. Mr. Cameron cheered as Messrs. Yorke and Greenwood worked their way through a set list of thirteen Radiohead songs, including one of his all-time favorite tunes, "Fake Plastic Trees."

That London gig was the first of a string of concerts around the United Kingdom, the pinnacle of a broad public campaign that drew in millions of ordinary Britons, who did much more than go to concerts. They marched. They wrote letters to members of Parliament. They enlisted their neighbors to do the same. The campaign—formally named the Big Ask—was devised by an environmental group, Friends of the Earth, in an attempt to get a serious law through Parliament to battle climate change. Mr. Yorke served as a spokesman throughout the campaign. Asked in an interview why he was doing it, he said, "It's good to get involved with something that's positive and constructive, rather than falling into the trap of thinking there's nothing you can do about climate change."

The year before the concert, Mr. Cameron had been elected as the young leader of Britain's Conservative Party, then out of power. He had won the post by promising "to switch a new generation on to Conservative ideas." Just four months after the concert, he made clear one of the ways he would do that: he stood with the head of Friends of the Earth and announced that the Conservatives would pass legislation on climate change if they got the chance. Mr. Cameron's announcement started a stampede of other politicians responding to the public's agitation for action. The Labour Party government, then in power under Tony Blair, hemmed and hawed at first, but ultimately came around. The Climate Change Act, the first law in the world to make solving climate change into national policy, passed Parliament in late 2008.

Two years later Mr. Cameron, aged forty-three and riding a wave of public approval for his position on climate change and his support for gay marriage, became the youngest British prime minister since 1812.

All those wind turbines in the North Sea did not get there by chance. Britain had been experimenting with offshore wind for nearly a decade before the Climate Change Act passed Parliament. But the bill led to increasingly robust national targets to cut emissions, especially from electricity production. Since clean electricity would be needed to displace fossil fuels in other sectors, the new Climate Change Committee declared, "radical reductions in emissions in this sector are essential if overall greenhouse gas targets are to be achieved." British politicians decided that offshore wind was one of the country's best options for achieving these goals. Successive candidates for the office of prime minister continued to raise the nation's aspirations, culminating in Boris Johnson's recent promise to build enough offshore turbines to power every home in Britain. It took years of government support before the offshore wind industry started down the learning curve. But now the industry is able to reward this escalating string of political commitments with falling wholesale power prices.

The United States has not passed any equivalent of the Climate Change Act, unfortunately. But the nation does have a history of helping to scale an important form of clean energy: *onshore* wind power, the very technology that got an important tryout in the California mountain passes of the 1970s. In the next chapter, we will describe how that happened. Here's a hint: just as for offshore wind in the United Kingdom, advocates had to use the tools of democracy to get onshore wind to scale.

In this chapter, we have discussed wind turbines and solar panels at length, but not because we believe they are the only solutions to the problem of global warming. In fact, they are only partial solutions. They can help us cut emissions on the electricity grid, but by themselves they will not get us all the way to a clean grid. They have their problems as sources of electrical power, as we will discuss in later chapters. And they do nothing to cut emissions from airplanes, or cargo ships, or the massive emissions from the production of virgin cement and steel. These are all problems that require other technologies.

We hope, instead, that you will see the scale-up of wind and solar power over the past twenty years as a paradigm for *how* we need to solve the rest of these problems. The theory of the learning curve provides some basic guidelines. For starters, asking what a clean-energy technology costs today is always the wrong question. If the technology is new, it is certain to be expensive compared to technologies that have been around far longer. The real question is: What will that technology cost in five years—or ten, or twenty—if we make a conscious decision to scale it up?

The stark reality is that scaling up a new energy technology requires somebody to pay a high price for it in the early years. If this sounds wrongheaded, remember that the same thing is true of consumer goods. In the 1990s, the market for flat-screen televisions began to scale on its own because everybody who saw one wanted one, and enough people could afford to pay the thousands of dollars that those early models cost. Consumer goods like cell phones and personal computers scaled for the same reason—the early adopters kicked those technologies onto the learning curve, driving down the cost to the point where most Americans could afford them.

What makes energy technologies especially complicated is that they are not governed by consumer demand in quite the same way. When you flip on a light switch, how much time do you spend thinking about how those kilowatt-hours are being generated? The public does not have much direct say in the way we generate electricity—but the government does. In fact, as we'll discuss in Chapter 2, for most of its history, electricity generation has been under government supervision. Consumers have more say over what kind of cars they buy, but even there, the government exercises enormous control—for instance, by setting efficiency standards that determine what kind of cars are allowed into the market in the first place, as well as exhaust standards to protect our air.

The decision to scale wind turbines and solar panels to the point where they became affordable required action by governments—a combination of carrots, in the form of subsidies, and sticks, in the form of legal requirements for the power mix. The success of those policies is one reason we are optimistic about the potential for change in the electric-power business. But at a global scale, power generation represents only 38 percent of carbon dioxide emissions. We have other immense problems to solve, and that means we must nurse new technologies along until they, too, can make a big contribution.

The rest of this book is about how we can do that. The government, all the way from town councils to Congress, exercises enormous influence over the technologies available in the marketplace. By inserting themselves into the most important of these decisions, ordinary citizens can help to capture the magic of the learning curve.

CHAPTER 2

Power Switch

The farm where Lola Spradley grew up, on the eastern plains of Colorado, was a place of violent and changeable weather. The sky sometimes dumped chunks of hail large enough to destroy her family's wheat crop, their main source of income. She was still a young girl in the 1950s when her mother made a remark that stuck with her.

The family had a few small oil wells, known as stripper wells, on their property. Her mother noted that the extra income from selling that small volume of oil was a godsend in years when crop prices were low or harvests were bad, allowing the family to pay their property taxes and hold on to the farm for another year. The lesson the young Lola took from her mother's remark was that economic diversity was a good thing, and could help people through hard times.

That belief, she said, explained why she reacted as she did when a group of environmental advocates came to her in 2003 asking for help. By then, Ms. Spradley was the most powerful woman in Colorado politics. She had risen through the legislature to become speaker of the Colorado House of Representatives that year, the first woman to hold the job. She was a staunch Republican, but had a reputation at the state capitol in Denver of being willing to listen to anybody. The advocates explained what they wanted: a modest requirement, written into state law, that Colorado's large utilities would have to start buying renewable energy for the state's power grid.

From her background, Ms. Spradley knew that farmers across the Great Plains had long depended on windmills to pump water. "I'm going: a windmill, making energy instead of pumping water? I'm not seeing why this couldn't happen," she recalled in an interview for this book. "And knowing that farmers live from year to year, a lot of them, I'm going, OK, this could be additional income."

She tried for two years to push the measure through the Colorado legislature, over opposition from coal-dependent power companies. She failed narrowly in the last days of each year's session. Environmental groups finally threw up their hands and, with her support, decided to take a ballot measure directly to the people.

By this point, several states around the country had begun passing laws requiring that a certain amount of the power on their electric grids come from renewable sources of energy. The idea had been born in Iowa, and it originally had nothing to do with wind or solar power. Instead, two entrepreneurs from Chicago were running around that state in the 1980s arguing that generators could be added to dozens of small dams, if the utility companies could be forced to buy the resulting power. They managed to get the ear of some politicians, as well as the Republican governor's office.

Among the politicians was a fellow named David Osterberg, a college professor and social and environmental activist who had won a seat in the Iowa legislature while living in a onetime chicken coop. He had moved into the shed to prove a point: "You can live in America on what people throw away." The county health department tried to force him to move out, resulting in a long legal battle that he won. He lived in the shed for five years, and he figures the publicity helped him win his legislative seat in 1982.

"I only won by four hundred votes," he recalled. "There were at least two hundred of these good old boys saying, 'I don't like professors, but that son of a bitch, he's been fighting the gov'ment, so he's got to be my guy.'"

As a freshman lawmaker, he took up the cause of the dams and managed to get laws passed to require the state's power companies to buy renewable energy, resulting in a long legal battle. By the time the electric companies finally lost, in the 1990s, the idea of dams had faded, but the companies complied with the mandate by buying power from a pair of wind farms. That was the beginning of a massive conversion of Iowa's power system to run primarily on wind.

"You had to have the policy," Mr. Osterberg recalled. "We had to say to those benighted people who ran the utility companies: try it, you might like it."

In the mid-1990s, the Iowa mandate caught the attention of a young woman in California, Nancy Rader, who had already done some deep thinking about how to incorporate renewable energy into power markets. A wave of reform was sweeping across the country to make those markets more competitive, but with no mechanism to attach any value to the environmental benefits of renewable energy. Ms. Rader and other advocates feared the small renewables industry of that era was about to get crushed.

She proposed an ingenious solution. She urged states to incorporate a renewables mandate into law, but with a twist. Generators of renewable energy would be allowed to issue certificates, one for each megawatt-hour they generated, meant to represent the "green-ness" of their method of power production. Retail electricity providers covered by a state renewables mandate would comply with it by buying these certificates—in effect, buying renewable power to pass along to their customers. The price of the certificates would be determined in a marketplace, creating competition among the renewable generators that would force them to find ways to cut costs. It was a clever marriage of mandates and markets.

When Ms. Rader first proposed it, regulators approved the idea, but the California legislature subsequently rejected it, only to come around several years later. A group called the Union of Concerned Scientists helped to push the concept nationwide. By the late 1990s, Massachusetts, Nevada, Wisconsin, and several other states had adopted variations of the requirement. So, remarkably, had Texas, with a focus on tapping the powerful winds in the western part of that state. The governor who signed the measure into law was a rising star of Republican politics: George W. Bush.

Utilities fought these laws in most places, but often not very hard—the requirements were typically modest, sometimes as low as 1.1 percent of electrical generation. The laws generally went through as part of the larger negotiation over reforming the power markets; environmental groups and the renewables industry extracted these modest targets as the price of their support.

The fight was tougher in Colorado, an oil- and coal-producing state where many jobs were tied to fossil fuels. After failing in the legislature, environmental groups there settled on a ballot measure requiring that large electric providers increase their purchase of renewable energy to a target of 3 percent by 2007, and 10 percent by 2015. The power companies argued that rates would go up as they bought more costly sources of energy, but the advocates wrote the measure to limit any increase to 50 cents per family per month. The opponents claimed that even modest levels of intermittent renewable energy might crash the electric grid.

Everybody knew the ballot measure was going to lose in rural Colorado and would probably win in the booming urban parts of the state. Ms. Spradley barnstormed rural counties in support of the measure, and her support was likely crucial to cutting the losing margin there. In an election in which Mr. Bush carried Colorado on his way to winning a second term in the White House, the voters also sent a strong signal about where they stood on clean energy. They passed the measure, Amendment 37, with nearly 54 percent of the vote.

That outcome resonated across the United States. If it could pass even in a coal-and-oil state like Colorado, perhaps that meant the American people really wanted cleaner energy. In pushing the states to adopt laws like this, environmental and renewables groups had a strategy. They well understood that renewable power was still costly compared to fossil fuels. But they had taken to heart the lesson of the learning curve: if you could drive up the deployment of clean energy, you would likely drive down the cost. And if it got cheaper, some of the resistance to it would go away. In the Colorado campaign, one advertisement declared explicitly that "the price of renewable energy will decrease as technologies improve."

In the case of wind energy, the rural economic development argument was proving to be powerful. While Congress under Ronald Reagan had allowed the tax breaks of the industry's early years to expire, it began in 1993 to offer a more modest tax advantage, paying the operators of wind farms 1.5 cents for every kilowatt-hour of power they produced. The champion of that measure was Chuck Grassley, the Republican senator from Iowa, where farmers stood to gain millions from the development of wind parks.

Eventually, thirty of the fifty American states plus Washington, D.C., passed binding clean-energy requirements, mostly in the late 1990s and early 2000s. The combination of state requirements and federal tax incentives drove the American wind industry toward significant scale. For several years, the United States was the world's largest market for wind turbines, though it would eventually be surpassed by China. And the learning curve worked its magic: every doubling of installed capacity led to a decline in costs of a little more than 10 percent. How did companies do it? The details should sound familiar by now: Factories got bigger and more efficient. Workers grew more skillful. The wind turbines themselves grew larger and taller, enabling them to capture more energy. Maintenance costs dropped as more reliable components were invented, and more parts were moved "down tower" where maintenance was easier. Dozens of innovations contributed to the steady price declines. If you happened to be one of the people who understood the theory of the learning curve, these technological improvements—and the falling costs they made possible—were exactly what you'd expected.

As we write these words, some small American states are getting 30 percent or more of their electricity from wind turbines. In 2020, nearly 60 percent of the power generated in Iowa came from turbines. Even in mighty Texas, the second-most populous state in the country, that figure is 20 percent. Nationwide, more than 8 percent of the nation's power is coming from wind farms, with plenty of room for them to grow—and room for the costs to fall further. They have already fallen so far, though, that wind turbines are the cheapest way to produce electricity across large swaths of the country. In early 2019, three-quarters of the coal-burning power plants in the United States were uneconomical just to *operate* compared to building new wind and solar farms from scratch. And the economics are changing rapidly; by 2025, more than 85 percent of the coal fleet is likely to be uneconomic. Driven in part by the boom in renewable energy and in part by cheap natural gas, well over half the coal-fired power plants in the country have already shut down.

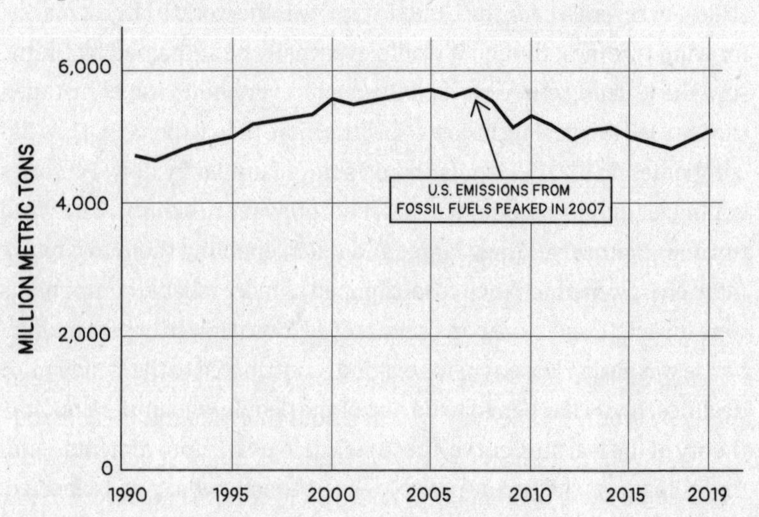

The polarized debate over climate change in Washington can be misleading. The truth is that in the middle of the country, where the big scale-up of wind energy has occurred, Republican politicians have played a critical role in supporting it. That is not because they are secretly worried about climate change; outlandishly, many of them still deny that it is a problem. But they see other benefits to renewable energy, especially rural economic development. Wind-power companies pay taxes to support local schools, they hire local people, and they make lease payments that flow into local economies. Ms. Spradley's first instinct in Colorado was correct: it really was good for the farmers. Sam Brownback, when he was the arch-conservative governor of Kansas, made this economic case explicitly. Mr. Brownback, a former national vice president of the Future Farmers of America, understood that the hefty payments for hosting turbines might be the salvation of some family farms. "We export lots of things, and in our future, I want us to export a lot of wind power," he declared in 2011. "We need more of it, and we need more of it now."

What the United States and some other countries did to scale up wind power, the Germans did for solar power. Starting in 2000, they passed a national policy to pay above-market rates for power from solar panels. They expected German panel producers to capture much of the benefit, but things did not turn out that way; the Chinese, seeing a potentially important new industry developing in Germany, rushed into action. Over the space of less than a decade, with billions in subsidies from their government, Chinese companies conquered the global market for solar panel manufacturing, even as that market exploded. It was as though the two countries had an unwritten pact to drive down the cost of solar power: Germany overpaid, and China overproduced. And once again, the learning curve worked its magic: if you go all the way back to 1976, the cost of solar panels has been falling by around 20 percent for each doubling of cumulative installations. If you look only at the past decade, when most of the world's solar panels were installed, that figure is closer to 40 percent, but it's possible that shift is temporary. To be cautious, you might forecast that continuing to scale the solar industry will drive the panel costs down by 20 or 30 percent with every doubling of the installed base. In particularly sunny parts of the world, solar panels are now the cheapest way human beings have ever found to produce electricity at scale. For years, $40 to $50 for each megawatt-hour was considered a good wholesale price for electricity. (That translates to 4 to 5 cents wholesale for each kilowatt-hour, the unit of power that appears on residential electric bills. American homeowners in the lower forty-eight states usually pay a retail rate of 10 to 20 cents, which includes the cost of transmitting and distributing the power, plus profits for investors.) In windy locations, wind turbines can now produce power for about $30 per megawatt-hour, and contracts have been signed in India and Mexico for solar farms at less than $20. The era seems not far off when solar panels will produce power for less than $10 per megawatt-hour—that is, less than a penny per kilowatt-hour, a price so low that people are talking about relocating electricity-hungry aluminum smelters to the desert. Already, as we will discuss later in the book, a giant mill in Colorado that uses electricity to melt down scrap steel is installing a massive solar farm next door to supply some of the power.

The huge price declines of recent years for wind and solar power are celebrated, and rightly so, but it is critical to keep in mind how long it took for these sources of power to become affordable. That is because they languished for so many decades, with no government and no power company trying hard enough to push them to scale. For newer clean-energy technologies under development today, we do not have half a century to dawdle. To stop scorching the Earth, we need to bring emissions of carbon dioxide pretty close to zero by around 2050. That is less than thirty years away. To meet that goal, potentially important new technologies need to be pushed from the laboratory into widespread use as quickly as possible. Now, understanding what it really takes to create affordable technologies, we can be realistic about what society must do in these precious decades just ahead of us.

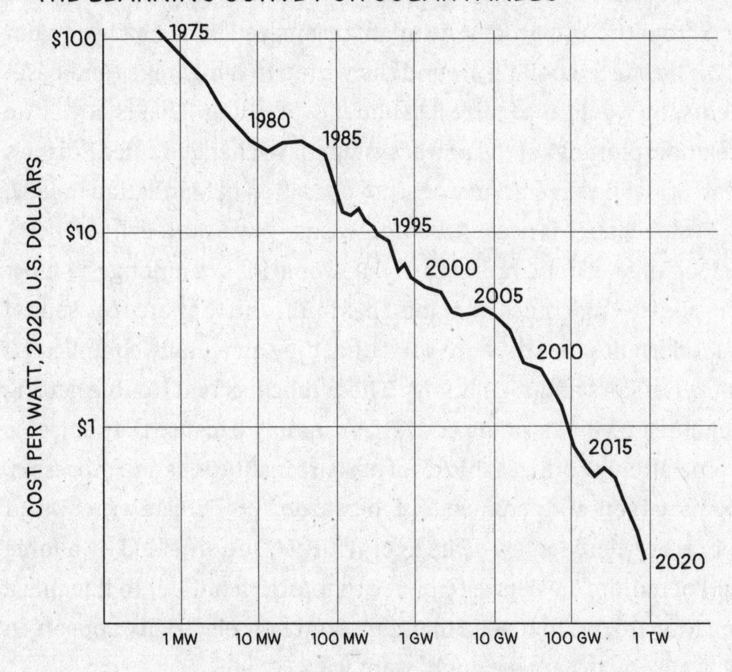

Cleaning up the power grid is the key to cutting emissions across the whole economy. That is because electricity can substitute, in many energy uses, for the burning of fuel. The basic strategy is called "electrify everything," from cooking to heating to transportation to factories. But that is only going to work if we eliminate emissions on the grid.

And there is good news: the spread of renewable energy and the advent of cheap natural gas have caused power emissions to fall sharply in the United States since they peaked in 2007. But we still have a long way to go. Across most of the country, utilities are holding back from embracing clean energy at the necessary scale—partly due to inertia, it is true, but also because the rules under which they operate are antiquated, as is much of the technology they use. If Thomas Edison, who invented the electric grid in the 1880s, came back from the grave today, he might be amazed by our gadgetry, but when he started poking around the wiring in our garages and basements, he would recognize just about everything. That is how little the underpinnings of the power system have changed since Edison's Pearl Street power station went into operation in Manhattan in 1882.

Faster change is needed, but for reasons physical, legal, and political, change is not coming easily. Power grids are among the most complicated machines humanity has built, and they are so essential to modern life that they are surrounded by an equally complicated thicket of laws and regulations. The public has real trouble getting a handle on what is going on. "That's mainly because the subject is excruciatingly boring, a thicket of obscure institutions and processes, opaque jargon, and acronyms out the wazoo," the climate writer David Roberts declared a few years ago. "Utilities are shielded by a force field of tedium." We have to pierce that force field! But to imagine a way forward, we first need to understand how electricity appears in the walls the instant we might want it.

A Modern Miracle

Electricity is so much a part of our lives that we take it for granted, forgetting just what a bizarre commodity it is. It has to be created, transmitted, sold, and consumed in the same instant. It is a devil to store. And yet it's the most useful form of energy, by far. It allows you to focus power exactly where you want it, when you want it, with no moving parts, no smoke, and no flames in your home. You can send the exact dosage of electricity you need to power your iPhone or your stereo, or the power company can shoot it down to the municipal works to pump water for your entire city. Indoor plumbing might seem like the most indispensable modern invention, but it only works because electricity is available to run the water and sewer pumps.

Cheap, invisible power is ready on tap, twenty-four hours a day, every day of the year. Like magic, it allows our puny bodies to shoot to the top of forty-story buildings without breaking a sweat. It moves water over mountain ranges, so that people can play golf in Phoenix or grow lush lawns in Los Angeles. It allows us to defy the seasons and be warm in Buffalo in January, cool in Dallas in July.

Most people have no idea where their electricity comes from—and why should they? It is buried in the walls, available in plugs every six feet. In advanced countries it is reliable, too—while in developing countries, one of the hallmarks of deep poverty is a lack of access to electricity. When people get it for the first time, their lives are transformed. A young girl who might have struggled in school suddenly has lights to study by at night. Getting electricity to the 770 million people who still do not have it is a moral imperative.

This magical stuff that powers every aspect of a modern economy is affordable, too, at least by the standards of the global middle class. In a country like the United States, a monthly electricity bill can cost less than a cell phone bill. Think about that for a second: you get light, heat, entertainment, cold beer, hot coffee, safe food, worldwide communication, and access to the accumulated knowledge of mankind, all for a few dollars a day. It has to be one of the best bargains ever. But even in the United States, some people do struggle to pay their power bills, of course—people who need lights at night as much as anybody. So keeping power rates reasonable is another moral imperative.

Yet for all the good it does, electricity, the fundamental technology of the modern age, also does great harm. Worldwide, 38 percent of the carbon dioxide that is overheating the climate comes from power plants. China now emits more CO_2 than any other country, and a rising demand for power, met by a massive boom in coal-fired plants, is the biggest reason. The United States has made the largest historical contribution to climate pollution and is still second overall in greenhouse gases injected into the atmosphere, with 33 percent of our carbon dioxide coming from power plants. And we shouldn't forget that, measuring emissions per person, the United States still pollutes far more than China.

We truly cannot live our modern lives without electricity—but if we keep producing it the old way, we're going to have a harder and harder time living as we're used to. Remember when we said electricity was cheap? That's only true if you ignore the costs of the health, climate, and environmental damage associated with electricity production. Even discounting the climate problem, we need clean electricity to displace the combustion of fossil fuels, which creates air pollution that is estimated to be responsible for roughly one out of six deaths worldwide.

Destroying the public's health while wrecking the planet is a pretty high price to pay for cold beer and lights at night. So as desperately as we need electricity, we just as desperately need to find better ways to produce it.

Where We Get Power

The electrical current in your walls consists of electrons, tiny subatomic charges that flow along the wires embedded there. Most of those electrons were propelled into motion by a device called a dynamo, also known as a generator. Some source of mechanical power turns the shaft of the dynamo, causing magnets inside to move across a copper coil, inducing a current in the wires. That current may have come hundreds of miles to reach you, but it travels along power lines at nearly the speed of light. As you cook dinner, the power illuminating your kitchen started inside a dynamo just a fraction of a second ago. These days, depending on where you live, some of the power might also have come from a solar panel, which converts sunlight to electricity with no moving parts, or from a turbine that captures the wind to turn a generator.

Modern dynamos are colossal, and it takes immense mechanical power to turn them. Water flowing through a conduit in a large dam can do it, by hitting a water wheel attached to the dynamo. This was one of the earliest ways of producing power in bulk, and it is still one of the most important sources of electricity worldwide. In 1950, at the end of America's great dam-building era, dams supplied nearly a third of the nation's power, though now electrical demand is much higher and the proportion has dropped to less than 10 percent. A few countries, like Norway and Costa Rica, are so blessed with water power that it supplies essentially all of their electricity.

Most countries are not like Norway, though, and most modern electricity production involves heat. Water is boiled into super-hot steam, which shoots through pipes at high pressure, turning a turbine attached to a dynamo. Even a nuclear power plant, as elaborate as it may look in photos, is really just a fancy way of boiling water. You can think of nuclear plants as giant tea kettles powered by an unusual source of heat: the splitting of uranium atoms.

Most of our electricity, however, comes from burning fossil fuels to turn water into steam. For many decades, the fuel of choice in the United States was coal, and that is still true in many large countries, including China and India. Coal—the fossilized remains of ancient plants—is dug from the ground, delivered to power plants by the trainload, and dumped into machines that work like giant pepper grinders. The ground coal is then blown into a furnace and burned in a mighty fire that can easily reach more than a thousand degrees Fahrenheit. That fire, in turn, boils the water to spin the turbine that turns the dynamo.

Most coal plants are huge. A typical unit built, say, forty years ago, will produce enough electricity to run a half-million or even a million homes at once. As the great fire consumes immense quantities of coal ground to the consistency of talcum powder, the carbon within the coal combines with oxygen in the air, and the power plant pours millions of tons of carbon dioxide out its smokestack every year. In essence, power plants are using the atmosphere you and I breathe as a free dumping ground for their waste.

These lumbering coal-fed power plants form the core of the power systems in many countries around the world. In the United States, natural gas plays a growing part in electricity generation, but coal plants still account for 20 percent. There is a reason the world is so dependent on coal: global reserves are big enough to power the world for centuries, and they are distributed everywhere from Indonesia to Australia, from Poland to South Africa, from Virginia to Wyoming.

While the fuel may be cheap, building coal-fired power plants is not, with new models running to several billion dollars each. And these two features—low costs to operate, but high costs for construction—have had a huge impact on the shape of our power system. For once a coal plant is built, the most economical thing to do is run it full-time, likely for many decades. It takes those decades to pay back the initial costs, and a coal plant is a wasted asset if it sits without running.

Coal-burning power plants are not that efficient. Well over half the energy in the burned coal goes up the chimney as waste heat. As little as a third of it is turned into electricity. This phenomenal waste is measured in dollars and pollution. Coal power plants do, however, run a very long time. With proper maintenance and the periodic replacement of critical parts, they can operate for as long as sixty to seventy years. This may sound like a benefit, but the longevity of the plants means that, without a conscious choice to retire them, their inefficiency and pollution will be with us long into the future. This is a big reason environmental advocates have tried so hard to stop new coal plants from being built; any we build now we'll be saddled with through most of the twenty-first century.

There's an additional physical challenge that drives our power grid's design. Unlike a product like wheat or soybeans, it is difficult and expensive to store electricity. This means power suppliers—whether running on coal, gas, or solar panels—must supply exactly as much electricity to the grid as users demand from it, second to second. The critical importance of this balancing act was demonstrated in February 2021, when freezing weather knocked many power plants in Texas offline. With supply unable to meet demand, grid operators had to black out large parts of the state at exactly the moment people needed electricity to run their heaters.

Why do supply and demand have to match so precisely? Millions of pieces of equipment are connected to the grid: your refrigerator, your dishwasher, sophisticated industrial machinery, the breathing machines and heart monitors that keep people alive. All of this modern equipment can tolerate slight variations in the power supply, but large irregularities would damage or destroy some machines.

As demand for power ramps up, the grid usually meets it by bringing more and more dynamos online. Sophisticated operations centers try to forecast power demand, looking days ahead at weather forecasts and other factors, but they also have to respond to sudden emergencies—any power plant can trip off with no notice, and that lost power has to be made up immediately. Some small power plants are kept running at all times as reserves, ready to inject power into the grid at a moment's notice. Other large plants can be turned on with ten or fifteen minutes' warning.

The system runs without much margin for error. "Brown-outs" are the industry's term for demand exceeding supply, and in a brown-out voltage drops across a utility's territory. Low voltages can ruin electronic equipment—on the grid, in your house, or in a hospital. So to avoid this damage, if one part of the grid fails, power equipment in neighboring sections will detect the fluctuation and may disconnect. In the worst case, the result is a cascading shutdown that can plunge entire states into darkness. In 2003, a software bug coupled with human error set off a sequence of events that caused blackouts across the Northeast, until 50 million people lost power. The system worked exactly as designed—most of the equipment shut itself off before it could be irreparably damaged. But getting it all running again took a week in some places.

In short, this is a system that—while huge, cumbersome, and costly—must be managed with the utmost precision. Power engineers meet these demands by developing an array of power plants with different characteristics, and then dispatching them to meet demand as cheaply as possible. Nuclear power plants and coal-fired power plants typically run all the time: they are not built to ramp up and down quickly, and their efficiency plummets at less than full power. You turn them on and run them, sometimes for years, powering them down only for scheduled maintenance.

To meet more variable demand, such as a heavy air-conditioning load on a hot afternoon, engineers turn to sophisticated natural gas turbines. These are physically much smaller than coal plants and more efficient, turning about half their fuel into power. Over the course of a day, these plants can be ramped up and down a couple of times. For very short, spiky peaks in demand, a less sophisticated type of gas turbine, called a "peaker," may be used. These are small and fairly cheap, but also inefficient and dirty, so engineers try to use them only when the need is dire.

Most of these familiar elements of the modern electric grid were in place by the latter decades of the twentieth century, when the power industry came under rising criticism for pollutants that were causing smog, mercury poisoning, acid rain, and other problems downwind of the plants. An immense environmental cleanup, costing hundreds of billions of dollars, was carried out under government orders to reduce these localized pollutants.

In the twenty-first century, the power industry is being buffeted by new economic forces and a new demand for change: to cut the greenhouse gas pollution that threatens to wreck our planet. The good news is that technology from the oil and gas industry has already delivered a short-term answer, of sorts. The bad news is that this answer isn't good enough.

The Gas Boom

Underlying large parts of the United States are thick layers of a black rock called shale. Geologists figured out in the nineteenth century that oil and natural gas permeated the rock, but decades of efforts to extract it proved fruitless. Wells would give off just a little bit of gas or oil, then quickly die, so that drilling them was never worth the money.

After the energy crises of the 1970s created deep concern about the country's dependence on foreign oil, a legendary Texas entrepreneur named George P. Mitchell, who had made his fortune with conventional oil and gas wells, spent decades trying to unlock the shales. The government encouraged his efforts with millions of dollars in grants and extensive technical assistance, even though everyone considered the scheme a bit of a long shot. Mr. Mitchell was eighty years old and suffering from prostate cancer when his company finally cracked the problem.

Injecting a combination of sand, water, and chemicals deep into the ground at high pressure shattered the shale, and the sand kept the new cracks open to release the fuels. The technique came to be known as hydraulic fracturing, or "fracking." Fracking was eventually mated with other technologies like three-dimensional underground mapping, as well as directional drilling, which allows drill bits to turn sideways once they've penetrated a productive seam. This conjunction of technology and innovation eventually produced a boom in American production of oil and natural gas. Fear of shortages and rising prices gave way just over a decade ago to a newfound abundance, and the price of natural gas in particular plummeted.

For power companies, the surfeit of cheap natural gas that has come their way over the past decade is proving to be both a blessing and, in some ways, a curse. While coal remains the cheapest way to produce electrical power in much of the world, that is no longer true in the United States. So much cheap gas is now available that power companies have ramped down their coal plants and ramped up their gas plants to carry more of the load.

Not long ago, coal commanded almost half the power market and gas about 20 percent. They've since flipped places, with gas supplying twice as much power as coal, and wind and solar seizing 10 percent of the market from fossil fuels. And at the same time this economic change was playing out, new environmental rules came in force to clean up the emissions from dirty coal plants. To keep operating, some of the old plants would have had to install expensive new equipment to clean pollutants like mercury from their exhaust. Seeing an opening, environmental groups went on the attack, mounting a nationwide campaign to shut down coal-burning plants.

This combination of economic and political factors has resulted in a massive abandonment of coal plants in the United States—just over two-thirds of them have retired, or announced plans to do so, in the past decade. The 185 plants that are left are typically the newer and more efficient ones. Entire states have gotten rid of most or all of their coal plants, and the four largest coal producers have gone through bankruptcy reorganization.

When you burn natural gas instead of coal for power, you cut carbon dioxide emissions nearly in half (coal is mostly carbon and burning it produces carbon dioxide, but natural gas molecules have four hydrogen atoms for every carbon atom; burning hydrogen produces only water). This massive switchover on the electricity grid has been a big factor in causing United States emissions of carbon dioxide to drop; in the power industry, emissions are down 40 percent since peaking in 2007.

The drop is not fast enough, but it is certainly progress. It comes with a pretty big downside, though: the production wells and pipelines for natural gas tend to leak enough to throw a significant amount of methane into the air, and it is a potent greenhouse gas in the short run. That cancels out some of the benefit of switching from coal to gas.

Cheap gas competes not just with coal plants, but with the nation's nuclear power stations. Wholesale power prices have fallen enough that many nuclear plants are in economic trouble. More than a dozen had closed by the end of 2021, with more likely to do so in coming years. The nation still has ninety-three operating plants, but a quarter of them may ultimately be at risk of shutting down on economic grounds. These plants emit no carbon dioxide as they operate, and if more of them shut down precipitously, they are likely to be replaced by gas-fired power plants. In that scenario, greenhouse emissions will go up.

To keep the low-carbon power from nuclear plants online, some states have already bailed out a handful of reactors, citing the climate risk of letting them shut down. In late 2021, Congress finally came to the rescue with a $6 billion package of subsidies to keep the plants open, though it is uncertain if that will be enough to save all of them. How fast these plants go offline really matters for the climate. For reasons we will discuss later, efforts to build new nuclear plants have been a disaster, so slowing down the retirement of old ones that can operate safely is a necessity.

In economic terms, the large supply of natural gas has been a short-term boon. But if we are going to clean up the power system, gas-fired plants must go away within a few decades. Unfortunately, this fact has yet to penetrate the minds of many executives in the power industry. They have proposed to build dozens of new gas-fired power plants across the United States. If they are completed, these plants could operate for much of this century, leaving the country unable to meet any reasonable goals for reducing its greenhouse emissions. On the other hand, if the nation were to get more serious about climate protection, these newer plants might wind up being shut down early, potentially costing electric customers huge amounts of money.

If we slash our fuel-burning as much as the situation seems to demand, will we still be able to operate a reliable electric grid?

Winds of Change

The forecast was sobering. The summer of 2021 had barely started when a large heat dome settled over the western United States. Temperatures in Portland, Oregon, and nearby towns were expected to soar far past a hundred degrees. For the power company in the region,

Portland General Electric, keeping the lights on—and air-conditioners humming—would be an immense challenge.

But the company had an ace up its sleeve. For years, it had been working on clever ways to manage demand on the grid. The power company's computers were tied into thousands of customer thermostats and could kick them up a few degrees if the grid was stressed. More than a hundred thousand customers had signed up to receive alerts at such times so they could cut back on their own, in return for a break on their power bills. Portland General Electric had enlisted an intermediary company to put switches on several thousand customer water heaters, and it had signed up big companies who agreed to have their power cut back when energy demand was high.

As temperatures soared to 115 degrees in late June, the company pulled out all the stops. Its computers sent out a radio signal across the Portland region; in dozens of apartment buildings, the switches atop water heaters received that signal and trimmed the energy use from the devices. The water in people's water heaters cooled down slightly. It was past 5 p.m. on a scorching day and few of them were taking hot showers, so it's unlikely many of them noticed, but anybody who did could have easily overriden the signal. Similarly, the company tweaked thermostats in homes and businesses to let the air get a bit warmer, cutting power demand. From 5 p.m. to 8 p.m., the company managed to suppress electrical demand by sixty-one megawatts, helping to keep the rest of the grid supplied with power. On the hottest day in Portland history, only a handful of small, local power outages occurred.

"Smart grid" is a term that baffles many Americans, but when energy experts use it, this is the kind of thing they mean. For many decades, the demand for electricity was treated as more or less a given. Electrical utilities mostly focused on supply, running their cheapest power plants all the time and then firing up the more costly ones as demand rose. Most customers paid fixed rates that bore little relationship to the cost of power hour-by-hour. That was a dumb grid, you might say.

From the beginnings of the power industry in the nineteenth century, people realized that you could in theory work on both ends of the equation, manipulating electrical demand as well as supply. By the mid-twentieth century, it had become commonplace for utilities to sign up, say, big industrial users who were willing to have their power shut off in a grid emergency. But for decades, this approach found fairly limited use, not only because the technology was clunky but because utilities liked the guaranteed profits they earned building new power plants.

Now we are entering an era when, in principle, the demand for electricity no longer has to be taken as a given. The spread of new electronic controls and improved communication methods has opened new approaches. Consider the kind of technology that may arrive in the near future. When you put your dishes in the dishwasher at night, you probably hit the button to run it, then perhaps unload the dishes the next morning. But what if your dishwasher were smart enough to talk to the electrical grid and wait for the cheapest and cleanest available power, which might come at three-thirty in the morning, when all the lights are off and everybody in town is asleep? The dishes would still be clean the next morning, but your power bill at the end of the month might be a little lower. Similarly, think about the day when you might have an electric car in the garage. Charging a car draws far more power than a dishwasher, of course. But most people don't care what time it charges, as long as it's ready to drive the next morning. What if your car could talk to the grid, negotiating a charging time based on the lowest price for electricity? The power company could save money by spreading out the demand across its most efficient power plants, and it might be willing to split some of the savings with you.

Now imagine a world where we have installed far more renewable energy on the grid—instead of 10 percent of our power coming from wind and solar plants, which is the figure in the United States today, maybe we get to 50 or 60 percent. Most people know the big problem with wind turbines and solar panels: they only operate when weather conditions are right. But they are highly desirable sources of electricity, given that their fuel is free and that they do not pollute, the way fuel-burning power plants do. That future electrical grid is likely to be one where the supply of power is more variable—more intermittent—than it used to be. What if the demand for power became more flexible, too, with your water heater and my thermostat and the schoolteacher's electric car all dancing in real time with the variations in power supply?

The experiment in Portland is still small, with tens of thousands of devices hooked up. Portland General Electric intends to go big with the approach, though. By 2025, it intends to use such methods to suppress power demand equivalent to the output of one gas-fired generating station. Across the whole Pacific Northwest, it's been calculated that if power companies connect enough smart devices, peak electrical demand can be suppressed sufficiently to avoid having to build two or three gas-fired power plants—or perhaps, to shut down existing dirty plants. And it can be done with little inconvenience to anyone.

Across the nation, cleaning up the power grid will likely involve hooking up hundreds of millions of devices in this way. It will also involve sending signals to electric cars and grid-sized batteries telling them when to charge themselves. A research outfit called the Brattle Group found that by 2030, an ambitious strategy could potentially cut the peak power demand in the United States by 20 percent, compared to what it would otherwise be in that year. But nobody really knows how far we can go with this idea unless we try harder.

Unfortunately, we are still a long way from reaching that future potential. The Portland-based utility is in the vanguard. Across most of the country, utilities are holding back from doing this kind of thing at any real scale. They are afraid of cutting into their sales, and most of them have not been given any marching orders that would encourage them to adopt more flexible approaches.

If we push hard enough, how far can we go with the combination of a more flexible grid and clean power? For two decades, estimates of how much renewable power the grid can take have crept up and up. Finally, in a landmark study in 2012, the National Renewable Energy Laboratory concluded that the United States as a whole could reliably get 80 percent of its electricity from clean sources, with wind and solar supplying as much as half the power.

It looked, though, like a power system of that sort, however technically feasible, would be costly. The years since have blown up the latter finding, as renewable-energy costs keep falling. More recent studies argue that as much as 90 percent of grid power can come from clean sources using today's technologies, and that may actually result in *cheaper* power than the dirty grid gives us. And by the time we have conquered 90 percent, we will have more options for the last 10 percent.

Getting to a 90 percent clean grid will require some big steps, however. Remember that while some states are getting 50 to 60 percent of their power from renewables already, the country as a whole has not reached that level. Counting power from dams, nuclear power stations, and wind and solar farms, the American grid is about 40 percent clean today. States with high levels of renewables are able to balance their grids partly by trading power with neighboring states that still depend largely on fossil fuels. Getting the whole of the United States to 90 percent clean will require better ways to manage electricity demand, as in the Oregon project. It will likely require new kinds of power lines that can move power efficiently across long distances, so that wind farms in Wyoming or Kansas might supply power to the millions of homes in Southern California or Chicago. Huge new wind farms will have to be built in the ocean. Much cheaper ways to store electricity will be needed, too, to smooth out daily swings in supply. Batteries are another technology that has jumped onto the learning curve, and they are now getting cheaper every year—cheap enough to use at a large scale on the electric grid—but we need to push that trend along. As we will discuss later in greater detail, the country needs a stepped-up research program, and solving the problems of a grid with a high level of renewable energy should be one of its top goals.

Meeting these targets will require rapid progress in the next few years. The nation needs to build up renewable power as fast as it can, and use that clean power to cut out coal and scale down natural gas. In fact, meeting the climate goals laid out by the Biden administration will require building renewable energy at least four to five times as fast as we have been doing recently. That is the overriding climate imperative of the next decade.

Who, you may be asking yourself, has the power to make this happen? The answer, in part, is you.

Pulling the Levers

One by one, the mothers paraded up to a microphone at a government hearing room in downtown Denver. They had come from around Colorado to speak up for their children. The setup was formal, a bit like a courtroom. The mothers were importuning certain government officers sitting on a dais at the front, and they had deeply personal fears to relate.

"Kids spend more time outdoors than adults, and they are more active," Jennifer Clanahan said into the mic. "My daughter is a perfect example. She's constantly skipping and jumping and bouncing and running." And then Ms. Clanahan drove home her point: "As she is bouncing around every day, I want to ensure the air she is taking in is clean air, because particle pollution can increase heart disease, lung cancer, and can interfere with the growth and the function of lungs." A bit later, Christine Brescoll put a sharper point on the argument: "It's unconscionable to subject our kids to the long-term health consequences of coal-fired power plant emissions when safe alternatives exist."

Clean air was not the only concern the mothers had that night. "My kids belong to the generation that will be dealing with many of the impacts of global warming—the increasingly violent and damaging storms, the drought and increasing wildfire danger, the flooding," Amee Daigle declared. "I worry about what their future looks like and hope that we can take serious action before it gets much worse."

Perhaps you have never imagined yourself taking political action to battle climate change or support clean energy. Many of the mothers in that Denver hearing room were taking their own first steps. The testimony from the women occurred on a cold February night in 2018, and by then, Colorado had come a long way from the days when environmental groups had to beg for 3 percent clean power. The mothers were there not to fight with the local power company, but to support it. Xcel Energy, the same company that had fought the ballot initiative in 2004, had turned around as it gained more experience with renewable energy. Under a leader named Ben Fowke, the eight-state utility eventually committed itself to a sweeping environmental cleanup. One aspect of it was under consideration that February night: a plan to shut down two coal-burning power plants near the Colorado city of Pueblo and replace them with clean energy. The voices of the mothers were heard: despite opposition from conservative state legislators, among others, the plan got approved, and new wind and solar farms are under construction in Colorado. Across its territory, Xcel is already getting nearly half its power from clean sources, and has committed to 100 percent clean by 2050. Of all the large electric utilities in the United States, an analysis from the Rocky Mountain Institute found that Xcel is the one whose plans are most aligned with the goals of the Paris Agreement.

But who, exactly, were the mothers pleading with up on that dais?

To explain that, we need to step back in time a bit. When the electricity business was first getting off the ground in the late nineteenth century, chaos ensued. Competing power companies were stringing multiple sets of wires, creating an unsafe mess. City and state governments had to step in to impose order; economically, it made sense to string only one set of wires. But that effectively turned the power companies into monopolies. How could they be restrained from jacking up prices and earning unwarranted profits?

An entire discipline of American law sprang up to deal with this situation. Ultimately, all fifty states created powerful boards to regulate the electric industry. They were given the legal authority to set rates, approve expenditures, determine what power plants were built, and decide exactly how much profit the electric companies could make.

The boards are known as public utility commissions, abbreviated as PUCs, or in a few states as public service commissions, PSCs. The names are a bit of a joke: unless some controversy erupts, like a big increase in power bills, the public is barely aware these boards exist. Yet they play a profound role in every state economy.

Remember the secret economic levers we were talking about earlier, the ones that determine behind the scenes the way our economy works, setting efficiency standards for cars and new buildings? The PUCs are the custodians of one of the biggest levers of all: overseeing the way your power company plans for future electric generation.

The boards operate a bit like courts, with formal legal procedures. But unlike courts, they are obligated to listen to the public and make decisions in the public interest. Sadly, though, they are heavily influenced by the utilities they regulate. Governors usually appoint the board members, and the utilities may use their campaign cash to sway a governor's decisions. The public utility commissioners also get some of their marching orders from a state's legislature, and utilities seek to exercise their influence there, too. In many states, profit-seeking utilities are the biggest contributors to legislative campaigns. They ply lawmakers with fancy restaurant meals, free hunting trips, and so on. Even when it is not criminal, this kind of soft corruption undermines the public interest. That is especially true with regard to emissions, when so many of these utilities have big money tied up in coal- and gas-burning power plants that need to be shut down.

Sometimes, the corruption gets much worse. In recent years, two huge utility bribery scandals played out in the neighboring states of

Ohio and Illinois. In Ohio, FBI agents showed up early one morning at the farm of the speaker of the Ohio General Assembly, arresting him on bribery charges. He was accused of taking $60 million from a big utility, First Energy, and related companies to jam through bailouts for several money-losing nuclear and coal plants. In Illinois, another big utility, Commonwealth Edison, was accused of offering make-work jobs for the friends of legislators to win favorable treatment.

Under a time-worn regulatory model, utility commissions routinely grant the companies they control permission to build new plants and other equipment, and then authorize new electricity rates that allow the utilities to earn a fixed profit margin on that gear. This arrangement is not an incentive for economic efficiency; it is an incentive to build as much new equipment as possible, then stick customers with the bill. Remember when we explained that coal plants cost far more to build than to run? Even with stiff competition from natural gas and renewable energy, utilities in some states are fighting to keep dirty old plants open so they can keep earning a return on their invested capital. They are doing so even in places where the cheapest thing for consumers would be to shut down the coal plants; the utility's incentives are simply not aligned with the best interests of its customers.

Because they have so much power over the electric grid, these utility boards need to hear from the concerned public much more often than they do. It's not necessary to master all the complicated rules to have an influence. You can just sign up for emails from a group in your state that advocates clean energy, like the Sierra Club, the Southern Alliance for Clean Energy, Western Resource Advocates, the Natural Resources Defense Council, Vote Solar, the Southern Environmental Law Center, Fresh Energy, Mountain Mamas, the Environmental & Climate Justice Program of the NAACP, and many others. Then, you can respond whenever they ask for public support. If you cannot get to the state capital and testify to the PUC in person, you can write a letter, or file an electronic comment. In most states, you can now testify via a video call. The louder the public voice in favor of clean energy, the more pressure the commissioners will feel to get your state moving on the issue.

Another political target is to help persuade your legislators to give the commissioners clear goals that they must meet. The original clean-energy goals that many states passed a couple of decades ago have largely been met, and the targets have expired. They need to be replaced with more ambitious measures, such as raising a state's target for clean power, if progress is to accelerate. This is another opportunity for you to get involved: advocacy groups are already pushing new measures in many states, but they need backing from more people. Find out what's happening in your state and speak up! Even if you only have time to make a single phone call or write a single email to a lawmaker, that will help. Mention it to your friends and neighbors and ask them to help, too.

New, ambitious clean-energy laws have already been adopted in some states. California, Colorado, Hawaii, New Jersey, New Mexico, New York, Oregon, Vermont, Washington, the District of Columbia, and Puerto Rico have all set bold cleanup targets for their power grids. If your state is not on that list, it needs to be. During elections, look closely at the positions candidates have put forward on clean energy, and vote for the ones calling for bold steps. A national group called the League of Conservation Voters has affiliates operating under various names in more than thirty states, and they track candidate positions on climate and clean energy.

As states adopt new policies, the goal cannot be just to clean up the power grid; we must not ignore the workers and communities that will be affected by these changes. Some workers may have spent decades mining coal or running a fossil fuel–fired plant, with few obvious career prospects if the energy transition throws them out of work. As dirty power plants shut down, workers there will need help. In some cases that may look like job retraining for new careers. Or it could mean direct payments to workers, paid for by the cost savings from shutting down dirtier, more expensive coal and gas power plants. The solutions will vary place to place—and the workers deserve a seat at the table to determine what works best for them.

The same is true for marginalized communities who have long suffered the most from the fossil-fuel economy. They are more likely than wealthy neighborhoods to be situated directly downwind from coal-fired power stations, for example. Our focus should not only be on shutting down these polluters, but on finding ways to ensure these communities are among the primary beneficiaries of the clean-energy economy. This could mean developing renewable-energy projects where communities own and share in the revenues. While signs of progress are being made across the country, much more work needs to be done to ensure these communities, too, are equal participants in the decision-making.

In Chicago, two low-income, primarily Hispanic neighborhoods called Little Village and Pilsen had been subjected to decades of emissions from a pair of especially dirty coal-burning power plants. Residents tried for a decade to get them shut down before the campaign finally achieved critical mass in 2011 and 2012. Greenpeace joined the drive, scaling the chimneys of the power plants to hang banners. Other national groups helped. So did local college students. The final push came, though, from the neighborhoods.

"At first we sent a letter to City Hall and requested a meeting, and nobody wanted to talk to us," recalled Kimberly Wasserman, a neighborhood organizer who led the effort after her son developed asthma. The residents refused to take that for an answer, and young people marched down to Chicago City Hall.

"Forty-one young people laid on the floor and zipped themselves up in body bags and put inhalers in their mouths," Ms. Wasserman recalled. "We got a phone call from the mayor's media office that basically yelled at me and said, 'you embarrassed the mayor, this is not appropriate.' And our response was, well, this coal power plant in our neighborhood is not appropriate."

The Fisk and Crawford coal plants shut down in 2012, removing two of Chicago's biggest sources of air pollution—and two of its biggest contributors to climate change. It is a prime example of what citizen action can achieve. And this is just a preview of what we can accomplish if only we grab hold of the secret levers of our economy and, together, pull ourselves toward a better future.

CHAPTER 3

Where We Live and Work

On a Sunday afternoon in August, the newsroom of a major metropolitan newspaper would normally be a languid place. But on a particular Sunday in 1992, the newsroom of the *Miami Herald*, Florida's largest newspaper, was in a frenzy. Reporters and editors had rushed into work. By that evening, the newspaper was hawking an unusual special edition on the streets of Miami.

"THE BIG ONE," a huge banner headline screamed from the top of the front page: "Hurricane of our nightmares is knocking at the front door."

Those fears would prove to be well founded. The storm wobbled just before landfall and came in south of Miami, passing over Homestead, Florida City, and other communities at the southern end of Dade County. It assaulted the region as a Category 5 storm, one of the most powerful ever to hit the United States. Two *Herald* reporters staying in a Florida City hotel wound up trapped in a bathroom, holding up a collapsing ceiling as the hotel blew apart around them. The powerful wind tore off roofs, demolished electrical transformers in a shower of sparks, and found its way inside building after building, ripping them to pieces. Tens of thousands of people huddled in terror that long night as their homes collapsed around them; about forty were killed. When the sky finally cleared on Monday, August 24, southern Dade County was in ruins. The storm destroyed or badly damaged close to eighty thousand homes and left a quarter-million people temporarily homeless.

The remarkable thing, though, was that some houses were still standing, with relatively minor damage. Entire neighborhoods came through the storm in decent shape, right across the street from neighborhoods in ruins. Investigative reporters soon figured out that this was due not just to the vagaries of the winds. The real culprit was the local building codes. It turned out they were being poorly enforced. To cut corners, some builders had failed to fasten roof joists with metal brackets known as hurricane straps, and building inspectors had let them get away with it. The investigations found that materials and construction techniques were often substandard, and that turned out to be true across the entire state, meaning other parts of Florida were at risk of suffering the same fate. The storm bankrupted several insurers. Others began pulling out of Florida, rates skyrocketed, and the state was forced to step in with a rescue plan. Florida adopted new laws requiring much tougher building codes and stricter enforcement.

Hurricane Andrew became the most dramatic illustration in modern times of the significance of building codes, and of the building inspectors who enforce them. Mundane as they may seem, they could hardly be more important. Those few houses left standing after the storm were the ones constructed by builders who had spent the time and money to do the work right. Most home buyers, then and now, take for granted that builders follow the rules. And yet on that terrible August night, people lived or died depending on how diligently their builder had done so.

Society at large depends on such rules. They go beyond building codes to include fire-safety codes, pollution rules, federal standards for the provision of safe drinking water, and many others. Our buildings are, generally, safe places to live and work in because these public standards, developed over many decades, have made them so. As late as the early twentieth century, entire cities sometimes caught fire and burned down; today that is nearly impossible, and the reason is that fire-safety codes prohibit poor construction techniques and materials. They also prohibit life-threatening practices such as locking emergency exits. Whenever you move into a house or apartment, you can usually expect that clean drinking water will arrive at your taps, and wastes will be safely flushed away. Electrical codes ensure that you are quite unlikely to be electrocuted. Even if you accidentally drop a hair dryer into a sink full of water and, without thinking, reach in to retrieve it, a safety device in the power outlet should trip and cut the current in time to save you from annihilation.

Making and enforcing these codes is a human enterprise, of course, and thus prone to failure. Earthquakes, floods, wildfires—and yes, hurricanes—will sometimes show the codes to be inadequate. In the worst case, they reveal cheating, and then scandal erupts, as it did in Dade County. But on the whole, public standards work more often than not. The occasional failures are a crucial part of the story; in the same way that airplane crashes lead to safer airplanes, the instances when public standards fail ultimately lead to improvements in those standards.

That steady, incremental progress is the most important feature of building codes, and the reason we need to enlist them in the battle against climate change. Buildings are one of the nation's largest sources of carbon dioxide emissions—in part because most of our electricity is used in buildings, but also because buildings burn fossil fuels directly. In many regions of the country, natural gas is the customary fuel for space heating and the hot water that flows from faucets. In some regions, homes are still heated with fuel oil or propane gas stored in tanks. When you count these fossil fuels along with electricity use, buildings are the source of one-third of our national carbon emissions. Meeting the 2050 climate goals will be impossible without cleaning up America's buildings, and the way to do it is to subject them to new rules.

We mentioned in the last chapter that the basic climate strategy we must pursue is to "electrify everything" while cleaning up the electric grid. That must include electrifying the space and water heating of our buildings. The expansion of the gas grid into new buildings has to stop. More than fifty cities in California and a handful of other states have adopted or proposed rules to ban or limit new gas hookups. But the gas industry is fighting back, running advertisements with gauzy pictures of blue flames to make gas look nostalgic and inviting, instead of like the huge climate problem that it is. "Grassroots" groups have popped up to try to stop the bans, but enterprising journalists have established that these are fake, funded by the gas industry, and run by its public-relations agents. Several conservative state governments have passed laws forbidding their cities to adopt gas bans. The politicians behind these laws claim that the gas bans limit "consumer choice," not mentioning that their campaign coffers are stuffed with money from the gas industry.

The tasks ahead of us are, in principle, straightforward: we need to tighten the shells of our buildings, old and new, and electrify everything that uses energy in those buildings, so that emissions fall as the electric grid gets cleaner. It may sound simple, but with the gas industry spending enormous sums to try to protect its market, the reality is that this is going to be a massive political fight. We think it is possible to win that fight and clean up our buildings; in some parts of the United States, this is already starting to happen.

Building for the Long Run

It was a sunny afternoon in the middle of winter, one of those crisp, beautiful days that make you understand why so many people want to live in Southern California. But nobody was living yet on Chasma Street.

Up and down a one-block stretch of the road, saws were buzzing. Hammers were pounding. The *thwack* of nail guns punctuated the air as workers put finishing touches on twelve new homes. They looked much like many others under construction in Orange County, where a decommissioned military base has given builders room to put up thousands of houses, expanding the vast reach of suburbia that already stretches south of Los Angeles. But these were some of the most advanced mass-market homes being built anywhere in the United States.

One modern feature was obvious to anyone standing in the middle of Chasma Street: Lennar Corporation, one of the nation's largest home builders, was putting solar panels on every house. The company had started doing so voluntarily in the middle of the last decade, but in 2020, it became the law in California that every new house must get solar panels. In other states where Lennar operates, the company includes the panels voluntarily.

The houses going up near the city of Irvine were state-of-the-art in ways less obvious than those rooftop panels, though. The most important features were hidden in the walls and windows, largely invisible to potential buyers, and there was nothing voluntary about them. The homes were being built to California's strict building codes, which means they had lots of insulation to keep them cool in summer and warm in winter, along with advanced window coatings and other features to save energy and water while improving comfort.

One of the men responsible for Lennar's decision to adopt green features like the solar panels was David J. Kaiserman, who was then a vice president at the company. On a visit to the Chasma Street construction site, he stalked up and down the block with a cell phone pressed to his ear, watching the delicate ballet of workers hoisting windows into frames, sealing leaks, and connecting the solar panels on each roof to electrical panels in the garage. At a model home nearby, Mr. Kaiserman showed off dazzling wireless features, including computer routines that set the living room up for movie watching by lowering window shades and switching on mood lighting. He showed off fancy stoves, big cabinets, and large—but highly efficient—refrigerators.

"Could you see yourself living in a house like this?" Mr. Kaiserman said with glee. "These are some of the most efficient homes in America, but you are not going to feel deprived."

The homes that Lennar is building in Irvine, and the codes under which all California builders must operate, embody a fundamental principle that needs to be applied everywhere in the world: *waste less*. Even in advanced economies where energy efficiency has been a mantra for decades, enormous amounts of energy are still being wasted in poorly constructed buildings. And because so much of our energy goes to heat, cool, and operate our buildings, every poorly constructed building we put up becomes a liability for the climate long into the future. It is imperative to stop putting up bad buildings. We also need to figure out how to patch up older buildings to make them less of a liability for our wallets and our climate.

Staying warm in the winter and cool in the summer is a matter of health and comfort, to be sure, but it is much more than that: it allows civilization to prosper in what would otherwise be intolerable climates. Montreal in the winter and Dubai in the summer are livable cities only because of heaters and air conditioners. The consequence of powering those devices with fossil fuels is that the more you heat your building, the more you heat the Earth, and the more you cool your building, the more you heat the Earth.

The energy use—or waste—in buildings depends on how they are built. Dramatic differences are driven by the choice of components and construction techniques. If you have a passing familiarity with home improvements, some of this may be obvious: windows with two panes of glass in a sandwich, known as double-glazed windows, leak less heat than single glazing. Insulation in your walls and ceiling keeps down your utility bills. More sophisticated technologies come into play, too: advanced window coatings, for example, can act as invisible insulation. These so-called "low-emissivity," or "low-e," windows have captured the bulk of the new-window market for American homes and businesses. You get all the benefits of glass without the energy costs of older windows. But even these windows still have room to improve.

Buildings have to withstand continual assault by the weather. They are exposed to cold and hot temperatures, to sun, rain, and snow, to day and night conditions, to dry air and humidity. They also have to survive door-slammers, shower-lovers, fridge-openers, bright-light-users, and every other sort of human predilection. The ideal building can handle this all with minimal energy demand. How is that possible?

The first requirement is simple: good insulation in the floor, walls, windows, and ceilings. If you had a perfectly insulated building, you could heat it to seventy-two degrees Fahrenheit and drop the building into Antarctica, and it would never cool down. Perfection is not achievable, of course, but today we know how to do much better than in the old days. When a new building is insulated and any remaining air gaps properly sealed, much of the energy problem has been solved before the building is occupied.

The second element of good construction is thermal mass—or, simply put, heavy, dense materials used inside the building. A tile floor or a stone wall absorbs heat when a building is warming up, and lets it off slowly when the building cools. Thermal mass creates comfort, especially in warmer climates: think of how nice it feels to walk into an old church or other stone building on a hot day. Thermal mass is a free comfort machine, with no moving parts, no maintenance required, and no fuel consumed.

The third element of good construction is to select quality windows and then situate them properly. With different coatings, windows can absorb solar energy, which you would want in chilly Minneapolis, or reflect it, for steamy Miami. It sounds a bit magical, but it really works thanks to nearly invisible coatings on the glass. Yet the issue is not just the windows themselves: the length of the roof overhang above them is also critical. For a building in the Northern Hemisphere, south-facing windows need an overhang of just the right length so that in the summer, when the sun is high in the sky, the overhang will cast the windows into shadow, keeping the building cooler. In the winter, by contrast, the sun will be low enough in the sky, and the overhang just short enough, that sunlight will stream through the windows, warming the building. This elementary principle is routinely violated in modern building design. But when it is taken into account, it can create free heating and cooling for the life of the building. When the building designer has sufficient flexibility, a building can even be oriented with most of the windows facing south, to maximize winter heat gain.

Some of these design precepts were figured out thousands of years ago by multiple civilizations: the ancient Greeks, the Babylonians, the Pueblo Indians of the American Southwest. Socrates, the Greek philosopher, gave advice on how to situate buildings for maximum comfort. Imperial Romans knew all about roof overhangs. But when we learned to feast on seemingly unlimited fossil fuels, these design rules were discarded as bothersome and unnecessary. In the mid-twentieth century, builders across the world threw up skyscrapers directly facing the sun, with single-paned glass, poorly sealed—creating what amounted to huge solar ovens and then compensating with equally huge air-conditioning systems. The United Nations headquarters on the east side of Manhattan, built in 1947, is a premier example of this wasteful thinking. Every one of the massive panes in that building functioned as a heater. A mile away, the Empire State Building, completed in the early 1930s, was pretty wasteful, too. Both skyscrapers have been retrofitted with modern equipment to slash their energy use, but there is no reason today not to get the design right from the start.

If a building is properly designed before construction begins, the incremental costs of making it energy efficient are trivial, typically adding just a few percent to the total cost. And that up-front investment will be paid back many times in energy savings over the life of the building. And yet, even in the 2020s when all these principles are well understood, too few buildings are being designed with energy efficiency in mind. Why not, if the benefits outweigh the costs?

Let us introduce you to a nerdy term that helps explain the disconnect: "split incentive." The people who design and build buildings almost never occupy them, which means they are not going to pay the utility bills. If the architect can save time by not thinking through solar heat loads or overhang angles, she will. If the developer can save money by skipping insulation, which is invisible to potential buyers, he may do just that. And if the builder is not fastidious about insulating the heating ducts, buried deep in the walls, who is to know? Unfortunately, the likely tenant is generally not consulted when these judgments are being made.

A lot of people seem to imagine that the energy problems in buildings can be solved by sticking solar panels on the roof, offsetting their energy use. After all, the price of these panels has plummeted, as we described earlier in the book. But in an apartment building the size of a small New York skyscraper, you simply cannot get enough solar panels up there to offset more than a small portion of the building's energy load. Panels can certainly offset more of the load in a suburban house—sometimes all of it, averaged over the course of a year. They can also put a big dent in the electricity use of a big-box commercial building like a Target or a Walmart or a grocery store, and they ought to be installed wherever possible. But inevitably, even a building with lots of solar panels on the roof will at times pull energy from a power grid that still uses fossil fuels, so the less energy we waste in the building, the better off we are.

New buildings need to be highly efficient from the moment they open. But if the people creating the buildings don't have enough incentives to outfit them for that, and if the eventual tenants don't know enough to demand it, the only way to achieve the goal is to impose public standards.

New Rules

The regulations developed over the past two centuries mean the buildings in our cities may no longer be firetraps, but they are still climate liabilities. If we are going to treat climate change as the emergency that it is, the time has come to assign a new job to the building departments that oversee construction in our cities and towns. These governmental departments need to become frontline agencies for preventing climate change. As it happens, this work has already begun, but like so much else having to do with the climate problem, it has not progressed rapidly enough. Too many states, counties, and cities are still woefully behind in using their building codes to tackle the climate crisis.

Here's another of those secret levers that determines how our society really works: across much of the world, building codes are updated on a three-year cycle. Sample documents, known as model or base codes, are drawn up by organizations stocked with representatives from the fields of architecture, mechanical engineering, refrigeration, and so forth. For years now, these model codes have included requirements for tight building shells, decent windows, and other measures to improve the efficiency of buildings.

But even though new sample codes become available every three years, many American states and cities are slow to adopt them. A half dozen states impose no building codes at all, but other states and cities are simply tardy. The federal stimulus bill that President Barack Obama got Congress to pass in 2009 included incentives for cities to bring their codes up to date, but now, more than a decade later, many are still stuck with the building codes they adopted in 2009 or 2010. Putting a stop to this foot-dragging needs to become a key focus for climate activism.

For the growing number of localities displaying real ambition about tackling the climate crisis, there's an exciting opportunity to be in the vanguard of the building efficiency movement by adopting advanced energy codes (sometimes called step codes or stretch codes). These are building codes that go beyond the current basic code to adopt more advanced energy measures—discouraging new gas hookups, for example, so that newly constructed buildings run entirely on electricity.

Cities and states also need to start using building codes and other measures to push for a particular type of equipment in new buildings:

heating systems and water heaters based on a technology called a heat pump. A heat pump is a device capable of moving heat from one place to another, or, more accurately, from one temperature zone to another. You already have one in your home: your refrigerator. The machinery sucks heat out of the interior of your fridge and pumps it into the room. The warm space behind your fridge is the flipside of the cool space inside it. But your refrigerator is limited in that it can only move the heat in one direction. If you have an air conditioner in your home or office, it is also a one-way heat pump.

The heat pumps we really need in buildings are more flexible. Nate Adams, a home-energy consultant in Ohio, has a pithy way of describing them. "Heat pumps are bisexual air conditioners," he said. "They go both ways."

These modern heat pumps can handle both the cooling and heating loads for a house. In the winter, the heat pump pulls heat from outside a home and discharges it inside. This is possible even when the outside temperature is quite cold, in the same way that your refrigerator manages to keep moving heat from the cold interior into a warm room. In the summer, a household heat pump can pull the heat from inside and pump it outside, functioning as an air conditioner. Just as in a refrigerator, the work is done by gases that pick up heat when they expand, and give it off when they condense.

Heat pumps used to be a bit dodgy, and many people still have the impression they do not work well. But the technology has improved markedly over the past decade. Newer heat pumps are far more efficient than older equipment, even in cold climates. An old electric heater turns one unit of electric energy into one unit of heat; a good heat pump can deliver three or even four units of heat for every unit of electricity. You read that right: because they are moving heat, not creating it, heat pumps can be 300 to 400 percent efficient. Water heaters and even clothes dryers on the market already operate on the same principles. In coming years, these devices need to become standard in every building, from single-family homes to the largest skyscrapers.

Home heating systems and water heaters based on heat pumps already have a particularly large toehold in the Southeast. They are sometimes sold as premium equipment, at high prices. But even at those elevated prices, they are often cost-effective in new construction because of the long-term energy savings. As cities start using their building codes to require heat pumps in new construction, we expect the market to expand and the cost of the equipment to fall—another example of the learning curve at work. And as the prices fall, heat pumps can capture more and more of the replacement market for equipment that is wearing out in older buildings.

Plugging In

We've just spent pages discussing the ways that buildings use energy for heating and cooling. But of course, that's not the only energy demand: inside buildings, people plug things in. Every single device in a building that draws electricity—refrigerators, air conditioners, televisions, lights, and so forth—can be efficient or wasteful. Is your flat-screen TV converting electricity into heat all day long, even when it's turned off? How old is your refrigerator? Do you have any clue how much energy it wastes?

We both grew up in the 1970s, in a world of shag carpets, bell-bottomed jeans, and snazzy kitchen appliances painted in bizarre colors like avocado and burnt orange. We didn't understand it then, but behind the doors of those avocado refrigerators, bad things were happening. As the tiny refrigerators of earlier decades gave way to larger units, with little attention paid to good insulation, the energy use of refrigerators soared.

As the energy crises of the 1970s led to new scrutiny of that excessive energy use, manufacturers began working to cut it. California led the push, but ultimately the federal government adopted legally binding standards for energy use, and it has gradually tightened them ever since. The result has been a revolution in the energy use of refrigerators, with new models today using 75 percent less power than those of the 1970s. That trend has saved tens of billions of dollars in electricity costs in the United States each year, not to mention the reduction in global-warming emissions. Since refrigerators keep food cold twenty-four hours a day, and since they can last for decades, the improvement pays off year after year. The fridge standards save a typical homeowner nearly $200 a year, on average.

Today, other countries have adopted the same approach, and the targets go far beyond refrigerators. Here's an example that most consumers rarely think about: Do you happen to have an old electronic device that uses a heavy power adapter? If so, put your hand on that adapter after it has been plugged into the wall for a while. Is it warm? If it is, then it is wasting energy—converting your home's electricity into heat, not just the five-volt or twelve-volt power the gadget needs to operate. For decades these black cubes, sold across the world by the billions, were little energy vampires, sucking—and wasting—electricity whether or not the device they were meant to power was even plugged in at the time.

In the late 1990s, a scientist named Alan Meier at the Lawrence Berkeley National Laboratory, in California, took on this problem. He had met the leader of a Silicon Valley company that had figured out how to redesign the power adapters. Early the next decade, he built a display to show how much power the older devices were wasting—you could push buttons on a board to see the energy waste from various gadgets. He got to show that display to George W. Bush when the president visited California.

"President Bush was a tech geek. He liked gadgets," Dr. Meier recalled. "So when they started the display, he started pushing buttons for twenty minutes and told his energy advisor, 'We need to do something about this.'"

Overall, the Bush administration compiled a terrible environmental record. But on this issue, Mr. Bush did the right thing. His administration committed the United States to supporting a worldwide cleanup of the power-supply problem. Many countries joined in, adopting national standards in a coordinated way that forced manufacturers to get rid of the heavy, wasteful power bricks. Today, if you buy an electronic gadget, the power supply is likely to be lightweight and highly efficient, using modern electronic circuitry to minimize the waste of energy.

More recently, this same kind of efficiency improvement has happened with lights. The incandescent type of light bulb that Thomas Edison created in 1879 was always wasteful, converting only 10 percent of the electricity into light and wasting the rest as heat. Anybody older than about twenty will remember that the things could easily get hot enough to burn your fingers. Fluorescent light bulbs, including the small curly kind that became available in the 1990s and could be screwed into regular light fixtures, were much more efficient, but many people found the light too harsh for use in homes. Technology has delivered a solution, in the form of the light-emitting diode, an electronic gadget that converts electricity into light using 86 percent less power than the old Edison bulbs.

LED bulbs are one of the world's most dramatic examples of the learning curve at work. While it had been possible for decades to make red and green light-emitting diodes, blue ones were needed to complete the color spectrum and allow bulbs to produce white light. But finding the right mixture of chemicals to produce blue LEDs proved exceedingly difficult. Two Japanese scientists and one Japanese American colleague finally solved the problem in the late 1980s and early 1990s, eventually winning the Nobel Prize for their work.

The first LED bulbs to come to market, in the early 2000s, were exceedingly expensive, but they found buyers, many of them motivated by the realization that the new bulbs would last twenty-five times longer than the old-fashioned kind. The market started to grow rapidly, and the magic of the learning curve kicked in: every time total production doubled, the price of the bulbs fell 18 percent. The

United States Congress, in a bill that Mr. Bush signed into law in 2007, finally mandated a switch for most types of bulbs. Energy use for lighting in American buildings has been falling for years now, and the trend is certain to continue. Nobody is worse off for it—we still have all the light we need, and when the longevity of the bulbs is factored in, we pay less.

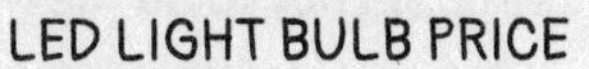

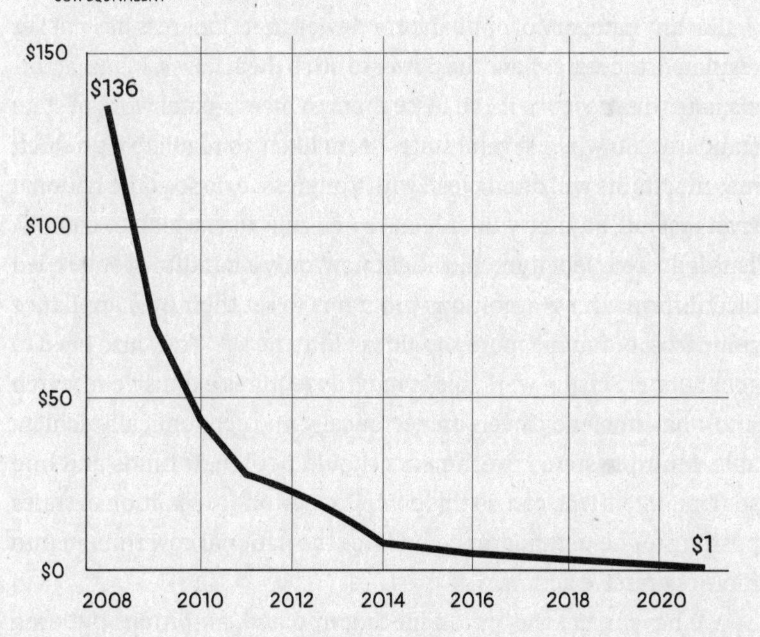

Refrigerators, power adapters, light bulbs—these are merely examples of the way we need to squeeze energy waste out of our lives and households. All our devices and appliances need to become hyper-efficient, delivering their services with the least energy possible. In a series of laws starting in 1975, Congress gave the United States Department of Energy a series of mandates for making new efficiency rules for whole categories of appliances, from dishwashers to microwave ovens to pool heaters. But how vigorously the agency pursues the mission seems to depend on whether a Republican or a Democrat is in charge of the White House. Republicans have usually been unwilling to push hard, and even Democratic presidents have sometimes dragged their feet. The agency has missed numerous legal deadlines for putting out updated standards. Congress needs to get much tougher with the Energy Department, penalizing it for blowing deadlines, and providing money for the agency to do the work required to meet them.

For any category of appliance or device that Congress has not yet regulated, the states have the power to do so themselves. Manufacturers hate this, because it can force them to meet a patchwork of state standards. So when several states seem likely to regulate a product, manufacturers will often plead with Congress to do so at the national level instead, so they will only have one rule with which to comply. This is an excellent dynamic! Right now, only a handful of states, led by California, have ambitious programs to set their own appliance standards, but many more should get into the act. They also need to collaborate. Setting workable standards requires extensive research into what efficiency levels are technically and economically achievable. Multiple state governments should pool their funds and hire staff members who can do this complex legwork. A coalition of states pushing for new standards could force the national government into more aggressive action.

We have made the argument that updated, ambitious building codes can cut energy waste in new buildings, and that aggressive appliance standards can cut waste by the gadgets that we plug into the walls. But if America is going to solve the climate problem, it will have to tackle an even bigger issue: cleaning up more than 100 million older buildings that waste energy and money. From modest homes to the tallest skyscrapers, just about every one is a problem that has to be solved.

Fixing Up

The first sign of trouble was the ice dams. When John and Sharon Poucher bought their Cape Cod–style home an hour outside of Cleveland, they figured it was well insulated. Certainly the energy bills were lower than the drafty old house they had left behind. But then one snowy winter, water started pooling behind dams of ice near the eaves of the house. It leaked through the ceiling. And it dribbled over the edge of the roof, freezing into icicles that hung off the house. Mr. Poucher, a skilled machinist by trade, knew enough about houses to realize this was not supposed to happen.

He tracked down a fellow we mentioned earlier in this chapter, Nate Adams, who has built a reputation as one of the most knowledgeable home-repair experts in the country—and who happened to live a few miles from the Pouchers. When he first heard from them, Mr. Adams knew immediately what was wrong: heat was leaking from inside the house through a poorly insulated roof, melting the snow. The water would refreeze near the colder eaves of the house, forming ice dams. Still more water would pile up behind the dams. It was the kind of problem that could really damage a home. Losing that much heat through the roof was also a waste of energy—and money.

But how big of a problem did the Pouchers really have? To figure that out, Mr. Adams dropped by their place one cold April morning. The couple watched curiously as he jammed a big fan into the opening of the front door, then sealed the space around it with red nylon fabric. The sunlight shining through Mr. Adams's rig cast an eerie Halloween glow on the upright piano in the living room. He set up pressure sensors, indoors and out, then turned on the fan. A terrific racket filled the house, and the Pouchers watched as Mr. Adams peered into every room, taking measurements with a handheld device.

By sucking air out of the house at high velocity, Mr. Adams had established a pressure gradient between the inside of the house and the outdoors. The indoor pressure drop caused air to come rushing in wherever it could. By comparing pressure readings, Mr. Adams confirmed what they had all suspected by now: the house had serious air leaks. The problem seemed at its worst upstairs in a small attic. So much air came rushing in that the Pouchers were astonished to see the carpet in that room float off the floor. Neither had Mr. Adams ever seen anything quite like it. He took a video to record the floating carpet for posterity.

With the blower still running, Mr. Adams moved again from room to room, snapping pictures with a special infrared camera. Warm walls would show up as orange splotches in his photos, but colder walls, where air was infiltrating the building, showed up as splashes of purple. To the Pouchers, the photos were vivid proof that their problems were real.

Over the coming months, they would do some hard thinking about whether they wanted to stay in a house with such problems. They looked at homes on the market, but could not find one they liked better. Ultimately, they swallowed hard and ordered up $60,000 worth of improvements, though only about half that work went to solve the home's energy problems. As Mr. Adams enlisted contractors to seal the air leaks, the Pouchers also replaced the roof of the house and built a big dormer window that turned the upstairs attic room into an elegant space.

And Mr. Adams talked them into his favorite upgrade: banishing their fuel-oil furnace and installing a heat pump. Mr. Poucher hesitated about that at first; he was aware of the bad reputation that heat pumps had in the United States, going back decades. But Mr. Adams assured the couple that the technology had come a long way and would keep them comfortable through a northeast Ohio winter. In deciding to fix up their house, the Pouchers were primarily aiming for comfort—but they were also happy when Mr. Adams explained the environmental benefits. By fully electrifying the heating and cooling load in their home, they would be joining the "electrify everything" movement, which meant the carbon emissions associated with their home would fall as the electric grid got cleaner.

John and Sharon Poucher are just one couple living in Ohio, but the kinds of issues Mr. Adams found in their house are pervasive across the United States. Codes and standards have been tightening over time, so the younger a house is, the more likely it is to be sealed properly. But Mr. Adams has repeatedly found problems even in newly built houses, a sign that the building codes are not always being properly enforced. And homes built in the middle decades of the twentieth century, in the great postwar suburban building boom, almost always have severe air leaks.

The issues go beyond single-family houses. Apartment buildings have problems with air leakage, too. Tenants have little control over this, even though in most of the country, they pay both the heating and cooling bills. This is a major issue for poor people, whose electric bills can be a daunting expense. And even large commercial buildings waste energy: offices, big stores, and the like in theory should have managers with the skills to cut their heating and cooling bills, and yet research has found extensive problems in such buildings. The air-handling systems were often poorly designed and installed, and even if they were installed correctly, knowledge about how to run them may have been lost over time.

The energy use of existing buildings is an immense challenge for the United States. It may ultimately prove harder to solve than the emissions from cars or factories. If we are going to meet the 2050 emissions goals, we have less than thirty years to stop burning gas in our buildings. We have to get tens of millions of homes switched over to electric heat pumps—but we also have to be careful that poor people are included in the switch, not stuck with spiraling bills for a rickety, declining gas system. Supplying the electricity needed to replace fossil fuels entirely from clean sources will be a major challenge. The more we can tighten the buildings and seal air leaks as we electrify them, the less clean power we will have to figure out how to generate.

Of course there is a private market for fixing up buildings in the United States, but it is far too small. Our energy bills might be higher than necessary, but most people just pay them, without doing the work to figure out what is wrong with their homes. To some extent, this impulse is foolish: repairs can sometimes cut energy bills enough to pay for themselves. But more often, achieving a truly comfortable house—and cutting greenhouse emissions to the maximum extent—costs more than can be repaid from the energy savings. Mr. Adams has found that selling his clients on major energy upgrades—"deep retrofits"—requires talking relatively little about the energy bills and talking more about the comfort of a home.

The United States must develop public policies that can break through the inertia surrounding this issue. This will be a delicate political problem, as forcing homeowners to spend large amounts of money against their will could induce a backlash strong enough to kill the whole effort. And this is where ordinary citizens come into the picture. If we expect our politicians to have the courage to tackle difficult problems like this, then we need to push them—and when the inevitable backlash begins, we need to stand with them.

Pulling the Levers

The critical first step is for city, county, and state governments to take ownership of this issue. Hundreds of American cities have nominally committed themselves to the climate goals of the Paris Agreement, yet many of them have yet to adopt the most recent building codes. And very few have come up with plans to repair their building stock over the next thirty years.

For citizens, the most immediate political target is straightforward: find out if your town has adopted a new code for energy use in buildings, something that would usually be done by the city council. If it has, that would most likely be based on the 2021 model code published by the International Code Council. If a city is lagging on this, then citizens need to demand that it be passed, now. And if local builders show up to fight the new code, they need to be countered by the voices of concerned citizens who understand what is at stake. In bigger towns, you may be able to find a local group that is already tracking this issue, or an environmental organization that would be willing to take it on. If not, think about starting one! With just a posting on Facebook or Nextdoor, you might be able to find like-minded citizens willing to go to City Hall and make the request.

If greater climate ambition is a possibility in your town, then the city council ought to consider going beyond the basic energy code published by the International Code Council. The ICC has published an ambitious appendix to the 2021 code that communities can adopt voluntarily, or they can look to an organization like the New Buildings Institute for help to develop a "stretch code" or "reach code." State governments can also encourage cities to adopt advanced codes, as California and Massachusetts have done. Every American state needs to be pressed to follow their example.

Once updated codes are in place for new buildings, the next step will be for city governments to set goals regarding the existing building stock. Only a handful of cities have done this so far, so we urgently need to push it everywhere. Cities must adopt building-repair laws whose stringency increases year by year. In major cities, the place to start is with large commercial buildings. City governments should adopt binding targets for emissions reductions, treating carbon emissions and energy waste in the same way they treat fire hazards and other threats to public safety. In some of the largest cities and in several states, this work has already begun. Colorado and Washington State, along with Washington, D.C., New York City, and St. Louis, recently adopted laws requiring the owners of large buildings to measure and report their energy use, and then to start reducing it. These governments are spending a lot of time holding the hands of building owners and managers. The likely effect of these laws will be to create a more robust private market in building repair. Similar policies are under discussion in several other cities and states.

Other American cities and towns need to copy these programs or find their own paths forward—so long as they recognize that voluntary goals are not enough. Targets must be backed up with fines or other penalties for building owners who refuse to get moving. Cities are perfectly willing to fine building owners for violating fire-safety codes; why shouldn't they be fined for scorching the planet?

To raise awareness among buyers of single-family homes, cities need to push more information into the marketplace. Incredibly, across most of the United States, houses' utility bills are not disclosed during a typical real-estate transaction. A buyer can often get that information by asking, but many do not. And yet it is a critical piece of information: unusually high utility bills in the depths of winter or the heat of summer will tell you off the bat that a house has a problem with air leakage. Ordinances requiring the disclosure of these bills ought to be adopted in every city and county across the United States, or by state legislatures. The nation's Realtors may well fight such requirements, but they would be unwise to do so: preliminary studies have shown that people are willing to pay several thousand dollars more for a house if its utility bills are disclosed. This appears to be true regardless of how high the bills are. That may seem strange, but it actually makes sense: if buyers know the figures, then they can do a better job calculating their likely monthly outlay on the house, which probably leads many of them to the realization they can offer a bit more to buy the property. So here is another political target for citizens: get a bill-disclosure ordinance on the agenda of your city or county commission.

The next step, after bill disclosure, would be to give prospective home buyers a sense of how a house compares to others in the city. A home's utility bills and square footage can easily be used to calculate a figure called energy use intensity, or EUI. Cities could require that owners be given a disclosure page that outlines the EUI and perhaps gives the home a letter grade comparing it to similar homes in the area. A home scoring a grade of A would be pretty tight by local standards, while a score of D or F would signal that major upgrades were needed. This measurement is not perfect, since the EUI of a home can be influenced by factors like how many people live there, but the limitations could be explained to homeowners.

If cities can spend the next few years creating much greater awareness about the energy consumption and carbon emissions of homes, they can set themselves up to enforce mandatory requirements starting in 2025 or so. Instead of a simple calculation like energy use intensity, the seller ought to be required to test the home and disclose its exact air leakage to prospective buyers. Eventually, local or state law ought to set a minimum efficiency requirement for a home that is about to change hands.

The method would be the same one Mr. Adams used when he aided the Pouchers: a blower-door test. This is a standard method of auditing a home, already widely available across the United States, and such a test needs to become a part of every real-estate transaction. The value of homes is appraised at the time of sale, so why not the energy use, too? In fact, since leaky homes waste money, the energy usage really ought to be factored into the appraisal of overall value. We envision local or state ordinances imposing a minimum requirement for home efficiency that would start fairly low and escalate over time. This means in the early years, only the leakiest homes would be prevented from being sold, and buyers and sellers would negotiate about who would pay to bring a home up to snuff, just as they negotiate today over leaky roofs or outdated plumbing. Over time, the standards should tighten, so that the entire housing stock of a community is gradually upgraded.

As we mentioned earlier, dozens of communities have already banned or discouraged gas hookups for new buildings, but those policies have to be extended to get older buildings off of gas. Cities need to find ways to promote—and later, to require—any furnace that gets replaced to be replaced with a heat pump. In the parts of the country where heat pumps are still unfamiliar, state governments could help enormously. We think they ought to intervene in the market directly—but temporarily—to shift the economics in favor of heat pumps. Many furnace and air-conditioner replacements are emergency repairs, carried out when an old unit breaks. States ought to impose a tax on the sale of gas-burning units, then use that money to offer rebates on heat pumps—rebates large enough to tilt a homeowner's economic decision in favor of the heat pump. As an extra incentive, states might also offer a cash grant to every repairman or -woman who replaces a gas furnace with a heat pump. A five-year program of this sort would alter the dynamics of the marketplace, tilting it toward heat pumps and weaning many buildings off the gas grid. A policy like this, if adopted by enough states, will help the heat-pump market in the United States achieve greater scale. And the magic of the learning curve tells us that it will drive equipment prices down, ultimately making the marketplace intervention unnecessary.

Across America, the work required to fix our buildings is enormous. Some taxpayer money will be needed, certainly, but it must be used selectively, to fix up housing for the poorest people. The bulk of the work needs to be financed in the private marketplace, and that means we need creative thinking. While many homeowners may ultimately be willing to do what the Pouchers did in Ohio—tap their savings or take out a home-equity loan—we think other financing methods will be required. One possibility is to copy a model that has worked in the rooftop solar-power industry. These days, you do not necessarily have to spend $30,000 up front for a solar system on your house. You can effectively lease it from a solar company, where you put nothing down and they pay the up-front cost. The homeowner pays for the clean power the panels produce, usually at a savings over the local utility's rates. We think a similar business model could work for home-energy upgrades, where the equipment is installed and then leased back to the homeowner. This basic idea—energy efficiency as a service—is already offered for both homes and businesses in some places, and needs to be adopted more widely.

As cities and states make their cleanup plans, they need to consider one more potential source of funding: electric utilities, which stand to be huge beneficiaries of the trend toward electrifying buildings. Many of them already run programs that offer homeowner grants for new, energy-efficient appliances. But these existing programs barely tap the vast economic potential. Imagine a fully electrified home with smart appliances, a sophisticated heat pump, an electric car in the garage, solar panels on the roof, and—perhaps—a house-sized battery to capture and store that solar energy for nighttime use, a battery of the sort that Tesla and other companies have already begun selling. Now imagine that all that equipment is connected to the local electric utility, which is allowed to exercise real-time control over how much power the gear is drawing from the grid. This would be similar to the situation we described in Chapter 2 regarding thermostats and electric water heaters, but now involving a much broader set of devices. Millions of houses wired like this would give the utility a tremendous ability to alter power demand on its system to match the variable power supply coming from clean sources like wind turbines and solar farms. Using the newfangled electric meters that are being installed across the country—"smart meters"—utilities would be able to calculate the exact value of this load-shifting. Such a smart system may allow them to put off expansions of the grid, potentially saving billions. To obtain this capability, utilities ought to be willing to put up a fair chunk of money to help homeowners accomplish the work they need to do. In a couple of dozen places around the country, they are already doing so. In the Sacramento Municipal Utility District in California, for example, homeowners who adopt a package of retrofits to electrify their homes can get rebates exceeding $10,000. You may be able to locate such rebates in your area by navigating to energystar.gov and searching "rebate finder."

In a few places around the country, utilities have started to offer even more ambitious home upgrade services to some of their customers. The idea is called Pay As You Save. The utility finances improvements to a customer's home that will cut energy use, and the customer pays back the cost as part of the monthly utility bill—but always winds up with a lower bill than before the upgrades were done. Customers of any income are eligible, but the programs are particularly useful for low-income consumers who might not otherwise be able to finance home upgrades. Rural electric cooperatives have been the biggest adopters of this idea so far, but for-profit utilities have started to test it. We think the state public utility commissions should push many more of them toward doing so—another political target for citizens.

Our goal needs to be nothing less than a nation of highly efficient, fully electrified homes and buildings in constant communication with an intelligent, clean electric grid. That world is still a long way off, and we will never get there until citizens start pushing their governments to make ambitious plans. The time to figure out how we are going to renovate America is now.

CHAPTER 4

Over a Barrel

Nobody really saw it coming—certainly not the Americans, whose vaunted intelligence agencies supposedly had eyes and ears all over the Middle East. When President Richard M. Nixon opened his Presidential Daily Brief on the morning of October 5, 1973, he could take reassurance from this sentence on the cover page: "Military exercises now going on in Egypt are larger and more realistic than previous ones, but the Israelis are not nervous."

They should have been. But ever since the Six Day War of 1967, the Israeli government had had supreme confidence in the country's military and intelligence services. That year, after Egypt closed an important strait to Israeli shipping, the Israelis had attacked and seized huge swaths of territory from their enemies, including the vast Sinai Peninsula. By 1973, Israel was dangerously overconfident, and missed the warning signs of an imminent Arab attack.

By the next morning, October 6, the country had lurched from blithe unconcern to panic. As Mr. Nixon's Presidential Daily Brief was being completed on that day, the Central Intelligence Agency slapped a late-breaking item onto the top page: Intelligence from Israel suggested "that Egypt and Syria are planning a coordinated attack across the Suez Canal and the Golan Heights before nightfall today." At 2 p.m., Egyptian planes started bombing Israeli fortifications, and within minutes, Egyptian ground troops had unleashed more than ten thousand shells on Israeli soldiers holding the far side of the Suez Canal. By nightfall, giant water cannons had blasted holes through an Israeli sand barrier, tanks and trucks were crossing the canal, and tens of thousands of Egyptian troops were pouring into contested territory.

In the north of Israel, just 170 tanks and 70 artillery pieces were in place to defend the country's heartland from five Syrian divisions with 1,400 tanks. To the south, in the Sinai, Egypt knocked out two-thirds of the Israeli tanks holding the peninsula within the first day of fighting. The rapid gains by the Arab armies left the Israeli public reeling. The war would prove to be Israel's closest brush with fate, the moment the Jewish state came nearest to being overrun by its enemies. Six agonizing days into the war, Israel was staring at the possibility of defeat. According to reporting decades later by the American journalist Elinor Burkett, the Israeli prime minister, Golda Meir, asked a friend to get her some pills "so that I can kill myself, so I won't have to fall into the hands of the Arabs."

The Israelis were running low on ammunition and equipment, and Nixon came under enormous pressure to resupply them. He hesitated: the Soviet Union was closely allied with the Arab states, and the war had the potential to become a confrontation between the two nuclear-armed superpowers. Mr. Nixon finally came to Israel's aid on the sixth day with an airlift to resupply the Israeli army and air force. As events unfolded, much of the American matériel would not be delivered until hostilities had ended, but the American commitment was a huge morale boost for the Israelis. After a credible showing by the Arab armies, the tide soon turned in Israel's favor. The conflict that Israelis call the Yom Kippur War, and that Arabs call the October War, would prove to be a tumultuous awakening—not just for Israelis, about the magnitude of the risks they faced from hostile Arab nations, but for the government and people of the United States.

Before October of 1973, only a handful of experts in this country understood how vulnerable the nation had become to a cutoff of foreign oil supplies. America had been unable to meet its own demand for oil since the late 1940s, but consumption was still rising briskly as Americans bought huge, gas-guzzling cars and took enthusiastically to their newly built Interstate Highway System. With only 5 percent of the world's population, the United States was sucking up nearly a third of its oil. By the early 1970s, economic bungling in Washington had worsened the situation. In the name of protecting consumers, the federal government was imposing price controls on oil, but that had the effect of discouraging new drilling and restricting domestic supply, making the United States more dependent on foreign oil. Even before the Yom Kippur War started, spot shortages of gasoline were occurring around the country.

Against this background of tight supply, lonely voices had been warning of the potential for an Arab embargo and the damage it could do. This was not some imagined threat; the Arab oil states had actually tried to impose an embargo during the Six Day War, but the United States at that point had spare production capacity to meet demand. America's own oil production peaked in 1970 and began to fall, and by 1973, global oil markets had little spare capacity.

Once it became clear the United States would go all-out to back Israel in the 1973 war, the oil states of the Middle East that were allied with Egypt, led by Saudi Arabia, began turning off the taps. Within weeks, the United States was in chaos. Gasoline shortages worsened quickly as a frightened public started hoarding supplies. As 1974 dawned, the tensions only intensified. Mile-long lines snaked from gas stations. People spent hours waiting to fill up, only to get to the front of the line and find the pumps dry. Fistfights broke out. Thieves siphoned gasoline out of other people's tanks in the middle of the night. Irrationally, the public blamed the oil companies for the crisis, convinced they were somehow holding back supplies to drive prices still higher. Angry truckers drove onto the interstates, came to a full stop with a hiss of air brakes, and switched off their engines, blockading the roads for hours at a time. Some truckers who did not support the strikes were killed by those who did.

An uneasy ceasefire was ultimately struck between Israel and its enemies, the embargo was lifted after five months, and Arab oil flowed once again. But, five years later, a revolution in Iran cut oil supplies once more, and Americans got to watch the ugly scene play out all over again—gas lines, public anger, the impotent flaying of the oil companies on Capitol Hill. The great oil shocks of the 1970s were defining events of modern history, forever altering strategic calculations in Washington and other world capitals. Both oil shocks drove prices higher, contributing to the runaway inflation of the 1970s that left a generation of Americans scarred by economic anxiety.

After the events of 1973 and 1979, America could no longer plead ignorance about the risks of being so dependent on oil. Experts came to understand that the cost of gasoline at the pump, displayed on foot-high letters all over the land, was basically a fiction; in the decades following the oil shocks, the American government would spend trillions of dollars and thousands of soldiers' lives fighting wars whose purpose, at least in part, was to keep the supply lines open from the Middle East.

With the perspective of nearly half a century, we can say that the nation has never truly taken the lessons of the oil shocks to heart. The United States is burning slightly more oil today than it did in 1973—less per person, but more overall. New technology is allowing the nation to produce more oil and import less, but the old habits of profligacy are basically intact. Whenever gasoline prices rise, Americans will switch to buying thriftier cars, and they will drive a bit less. When prices fall, the pain is quickly forgotten and sales of gas-guzzlers, like the sport-utility vehicles now so common on American roads, zoom again.

Despite that undying American predilection for wasteful vehicles, the situation we confront today is markedly different from the circumstances of the 1970s oil shocks. For the entire century that transportation has depended on supplies of oil, people have worried intermittently about running out. That fear seemed to come true in the 1970s—and again in 2008, when tight supply and rapidly growing Chinese demand briefly drove oil prices to a shocking $147 a barrel, double the level of a year earlier.

Now we understand, though, that the real risk is not that the world has too little oil. If that were true, we could count on shortages and rising prices to force some kind of transition to alternative vehicles. The real risk is that the world has too much oil, more than it can afford to burn.

Out the Tailpipe

These days, many people are excited about the possibility of a switch to electric cars. So are we. But cars last a long time, and the turnover of the fleet is slow. Many hurdles, including a lack of charging stations, stand in the way of full electrification of the world's automobiles. Even if we push hard, the transition to an electric fleet will take decades. That means as many as 2 billion additional fuel-burning cars will be sold before the changeover is complete. How efficient they will be is a critical issue in the near term—so we begin our discussion about how to fix the world's transportation system not with electric cars, but with the ones that still burn gasoline.

When you fill your car with gas, you undoubtedly think the point is to get it to move. You are right, of course, but the gasoline you burn mostly fails to do the job for which you bought it. Depending on the car, 75 to 80 percent of the energy in that gas goes to waste, as heat that dissipates uselessly from your radiator or through your tailpipe. As little as 20 percent of the energy is captured to turn the wheels and move the car.

From an engineer's perspective, this is—or ought to be—a professional embarrassment. One would like to think that most of the energy in a system would be used, rather than wasted! The waste is not only expensive for drivers; it means that most of the pollution spewing from cars, including greenhouse gases, achieves no useful work before entering the atmosphere. Here we are, more than a century after the Model T was introduced, and we are spending immense sums to pump oil from the ground, refine it, and then waste the vast bulk of it. That is no way to run a car, or a planet.

Is the problem fixable? With conventional gasoline engines, there are paths to improvement, but these are incremental. Energy transformations, such as chemical into flame, and then flame into motion, generate heat, and the more heat, the more energy wasted. That's why cars have radiators and cooling systems: they need to dissipate that heat—or it will literally melt the engine. The same thing is true in power plants, jet engines, or any other technology that converts fuel into motion. There will be no dramatic improvements in internal combustion engines, but steady improvement is both feasible and necessary.

How do we know cars can be made to be more efficient? Because it has been done before. After the first oil shock of the 1970s, a newly installed president, Gerald Ford, asked Congress to adopt mandatory fuel-efficiency standards for cars. President Ford called for doubling the mileage performance of new cars within a decade. The automakers claimed it was going to be impossible, that the government would put them out of business. But the law passed Congress, and the auto companies proceeded to do as required. Not only did the law require it, but the market at the time was pulling them hard in the same direction, as Americans fretted about high gasoline prices and looked for economical cars. This was the era in which the Japanese automakers, with their efficient small cars, broke into the market, permanently capturing a large slice of it from the Big Three auto companies in Detroit.

Engines were redesigned. Cars lost weight. The bodies were tweaked to slice through the air with less resistance. Each of those changes was incremental, gaining a mile per gallon here or three miles there, but together they added up to something big. To be sure, that era featured some spectacular failures. Several manufacturers tried to make small cars using quick-and-dirty designs, and we were treated to turkeys like the Ford Pinto, the American Motors Pacer, the Chevrolet Vega, and the AMC Gremlin. The idea of the Pacer was to make it as wide as a Cadillac, but shorter. The Gremlin looked as though a designer took a chainsaw to a normal 1970s vehicle. These cars were a laughingstock with the public, and they helped to give smaller American-made cars a bad name that they have not entirely shaken to this day. But the design sins aside, for more than a decade automakers focused intently on fuel economy, and by 1982, the average new car used about half as much gasoline as a new car sold in 1973.

In retrospect, though, President Ford and the lawmakers of that era made a critical mistake. Perhaps assuming that a future Congress and president would revisit the issue, they passed fuel-economy requirements that stopped ratcheting upward in 1985.

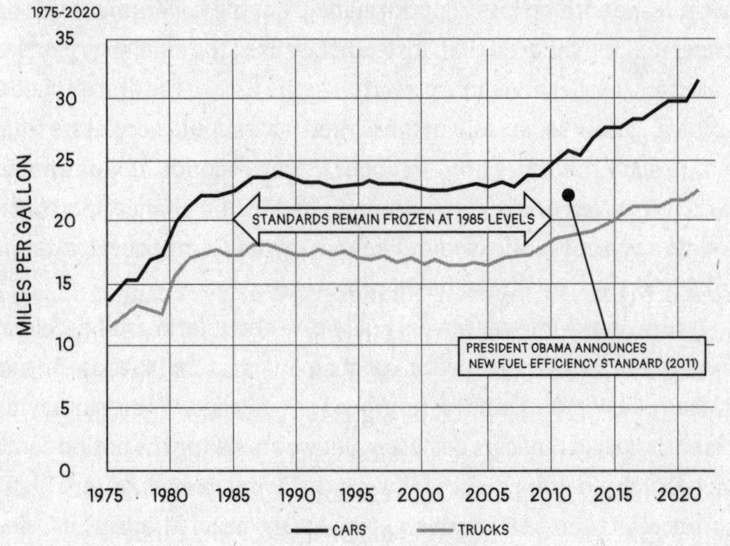

With the world's largest oil market running a far more efficient fleet of cars, and with new supply coming into the market from offshore wells in the Gulf of Mexico, from the North Sea and from new Saudi fields, the shortages of the 1970s gave way to bounty. The American president for most of the 1980s, Ronald Reagan, thus felt little pressure on gasoline prices, and had no interest in imposing tougher requirements on the car companies.

For twenty-five long years, the efficiency requirements stagnated. During the booming economy of the late 1990s, Americans started switching from cars to "sport-utility vehicles" or SUVs, which were technically trucks and subject to even weaker requirements. They were more profitable for the automobile companies, which marketed them heavily. Gasoline was cheap again, the average American seemed to be thinking, so why not buy a three-ton hunk of metal to haul your kids around to soccer practice?

The interesting thing is that automotive progress did not stop in those years—quite the contrary. Engineers continued to tinker, and engines steadily grew more and more efficient. In theory, the automakers could have used these improvements to give people better gas mileage with no loss of performance. But the government was no longer making them do that. Instead, they used the efficiency gains to make cars heavier and more powerful. At the time of the first oil shock, a typical family sedan sold in the United States could accelerate from zero to sixty miles per hour in about fifteen seconds. If you wanted to do it in seven or eight seconds, you had to buy a fancy sports car. Today, a typical family sedan like the Toyota Camry accelerates as quickly as the Corvette of half a century ago.

It turns out some car buyers really care about jackrabbit acceleration. As their engines got better, car companies catered to those buyers instead of using the gains for improved gas mileage. We are not saying it is entirely bad that cars got faster, but we are saying the nation could have struck a better balance between thrift and *vrooom*. After all, the extra acceleration generally just gets you to the next red light a bit faster.

The real tragedy—and lesson—is that the Congress of the 1970s did not write into law a standard requiring continual efficiency improvement. If the carmakers had been called upon to improve the efficiency of their fleets every year for the last half century, America would be using far less oil today than it is—and probably less oil than it was using in 1973, despite a 56 percent increase in the population. We might well have avoided being drawn so deeply into the fractious politics of the Middle East.

Fighting Over Money

When car companies redesign their vehicles every few years, they confront hundreds of trade-offs between power, weight, cost, efficiency, safety, and pizazz. Engine technology, in particular, is a high art. Today we have entered a sort of golden age in which computerized engine control and advanced materials are permitting some big gains.

How big? When the Toyota Motor Corporation introduced a redesigned Camry for the 2018 model year, it decided to employ a half-dozen new technologies at once. The result was an improvement in gas mileage of more than 20 percent in a single year. That kind of one-year increase is unusual, but it does illustrate how much progress can still be made. The Camry is not some hobbled, ugly car. It is the bestselling sedan in the United States, manufactured by a cadre of proud Kentucky autoworkers.

What are the ways that engines can be improved? The possibilities are numerous, actually. Increase the compression of gasoline and air before you fire it, and efficiency increases. Reduce internal friction through better bearings; adjust the timing of the valves to match power requirements; do the same with the engine displacement; compress the air entering the engine; coat the cylinders with low-friction materials. Those all sound pretty technical, we know, but the list gives you a sense of the available engineering options. It runs on to a hundred items. The catch is that each requires sophisticated design, and many of the changes add to the cost of producing the car.

That extra cost, not the technological challenge, is at the heart of the never-ending fight over automotive efficiency. Mind you, more efficient cars are highly likely to save people money over the lifetime of the vehicle—just as with efficient houses, the extra up-front cost is more than offset by the savings on fuel. But most buyers focus on the sticker price, not the differences in lifetime running costs, and with car companies fighting hard for market share, the extra sticker cost is a sensitive issue.

We believe that fuel efficiency can be doubled again, just as it was from the late 1970s to the mid-1980s. President Barack Obama set out to do just that early in his administration, increasing the standards from 27.5 to 54.5 mpg. As Gerald Ford's transportation agency had done, Obama's Environmental Protection Agency gave the automobile companies a decade to pull that off, ending in 2025. Those standards were good consumer policy, good energy policy, and good climate policy. They were even endorsed by the auto industry, which knew it had to do better on the issues of pollution and climate change. But the Donald J. Trump administration weakened the rules. Now the pendulum has swung again, with the administration of President Biden moving to toughen them.

We would argue that even the Obama targets were modest compared to what is possible. New engine designs are already proving to be 20 percent more efficient than any on the market, and close to 40 percent better than the average car on the road. They are designed to reduce waste heat, reduce friction, and increase combustion efficiency.

The engine is important, but so are all the other systems in a car. What happens when you turn on the lights, the sound system, and especially the air conditioner? The engine now has to provide power for those uses, too. Such accessory loads can be huge, but better design can minimize them. New paints are becoming available that look exactly like the old ones but reflect more of the sun's energy and cut heat gain in a car. Better glass can reflect some of the sun's rays back toward the sky, also cutting heat gain. On a July afternoon in Phoenix, that translates into less gasoline use for the air conditioner when the car is running.

Aerodynamics are becoming increasingly important. How slippery is the car, as it shoots down the road at seventy-five miles per hour? Are you pushing a huge volume of air, or slicing neatly through it? Look under a car: you will find a visual jumble of steering rods, shocks, exhaust pipes, drive lines, gas tanks, and more. Look under a plane: it is smooth. Perhaps cars will never be as aerodynamic as airplanes, but they have room to improve.

One of the biggest opportunities is to make vehicles lighter—and shrinking them is not the only way to do that. Today, the average car weighs more than 3,600 pounds, and SUVs are even heavier. The Lincoln Navigator clocks in at 6,000 pounds. That huge mass has to be stopped and started dozens of times on every trip, each time requiring energy. Advanced materials, like super-strong aluminum alloys, or carbon fiber, can slash weight without compromising safety—though they do, again, cost several hundred dollars more at the outset than old-fashioned sheet steel. The short-term focus on sticker price is a long-term disaster for the economy, for consumer pocketbooks, and for the environment.

If the car companies need to be told to do better, who should tell them? As you might imagine, setting efficiency standards for cars is a national responsibility in most countries. Only a handful of countries build the vast majority of cars, so a few of the world's governments have inordinate power over the global industry. And, we are sad to say, many of those governments are overly beholden to the car companies. Germany is the textbook example, where the Volkswagen conglomerate and other German car companies are so politically powerful that they have gotten away with scandalous behavior. The shocking discovery of a few years ago, in which Volkswagen was caught cheating on emissions standards, was only the most egregious recent example.

To some readers, our focus on making gasoline cars more efficient might seem narrow, even unimaginative, given the dawning possibility that we might be able to replace them entirely with electric cars. But here is the thing: we already have more than a billion fuel-burning cars on the planet, and that number is growing rapidly in the developing world. At a global scale, cars account for only about 9 percent of carbon dioxide emissions, but in post-industrial economies like the United States, they constitute a bigger share of the problem than the global numbers would suggest. With the ongoing cleanup of the electric grid, which has lowered emissions in that sector, transportation has become America's single biggest source of greenhouse emissions. We have nearly one car per adult in this country; many of our cars are gas hogs, and so cars and light trucks account for 22 percent of American carbon dioxide emissions. Add in heavy trucks and buses, and the ground transportation sector is responsible for 31 percent.

The rest of the world, including China's 1.4 billion people, seems to want to drive the way Americans do. And that is a problem, because the booming global car market could easily become a far bigger source of emissions than it is today. We think it is crucial in the near term to keep fuel use from growing as fast as the fleet is growing, and the only way to do that is with aggressive requirements to make cars more efficient.

Many of the issues we outlined above apply not just to cars, but to trucks and buses, too. They have been subject to weak regulations for decades, and while the owners of these commercial vehicles are more attentive to efficiency than the average car buyer, and willing to pay extra for it, the nation's heavy-duty fleet is still nowhere near as efficient as it could be if America were more focused on the problem.

The pollution from large vehicles is not just a climate issue; it is a health issue. Many heavy-duty vehicles run on diesel, and the fumes are a major source of air pollution. Students riding to school in diesel-burning buses are subjected to those fumes, at a time when their lungs are still developing and irritants in the air can trigger the onset of asthma. Poor children are at particular risk, because they are more likely to live near freeways or industrial facilities that send clouds of toxins wafting across their neighborhoods. For decades municipal buses were also a major pollution problem, triggering extensive efforts to convert them to run on natural gas or propane, but the cleanup is far from complete.

For cars, for SUVs, for buses, for urban delivery trucks, for long-haul big rigs out on the freeway, the near term strategy is the same: we need to make them more efficient. But longer term, we have to replace them all. The internal combustion engine, one of the defining technologies of the twentieth century, must go to its grave in the twenty-first.

An Electrifying Future

Fuel-burning cars were invented in the nineteenth century and became widespread in the early decades of the twentieth. But back then, if you really wanted to travel around town in style, you did not buy one of those smelly, noisy things running on gasoline. Instead, you bought a handsome, well-appointed electric car. You might have chosen a model from a fellow named Ransom Eli Olds, who dabbled in electric cars before switching to gasoline and creating the Oldsmobile line. Thomas Edison owned an electric car made by Studebaker. When President William McKinley was shot in Buffalo in 1901, he was rushed to the hospital in an electric ambulance, presumably the fastest conveyance at hand. (He died anyway, days later, of gangrene.) Cars of all kinds were still toys for rich folks in those days, but electric cars were popular enough at the turn of the century to outsell those burning gasoline.

Eventually, of course, the superior energy density of gasoline—which meant greater range for drivers—won out. But throughout the twentieth century, interest in electric cars never really died. As cities became more and more polluted, electric motors were seen as one of the most promising ways to clean up the air, particularly in California, which used its regulatory muscle to coax General Motors to bring a car called the EV-1 to market in the 1990s. But GM claimed there was little consumer interest, and managed to kill the EV-1 after only eleven hundred copies were produced.

Similar experimentation happened in other countries in the late twentieth century. In Norway, a small environmental group wanted to see electric cars get off the ground, and managed to interest a popular New Wave band—*a*-h*a*—in the cause. The group imported an electric car from Switzerland, but became exasperated by the bureaucratic runaround of trying to get it legalized for street use in Norway. They protested by driving around the country in the car, refusing to pay tolls and creating a commotion. The stunt got the attention of the government, which decided that maybe the kids were right—electric cars should get a tax break and preferential treatment for highway tolls and access to fast lanes. What could it hurt, the government thought, since only a few electric cars were being produced? It would prove to be a fateful decision in the long run.

What electric cars, including the EV-1, always lacked was a battery with enough range so that drivers wouldn't fear being stranded, powerless, on some lonely stretch of road—a worry that has lately come to be known as "range anxiety." That finally began to change by the early twenty-first century, thanks to a device first invented for use in power-hungry laptop computers: the lithium-ion battery. As those batteries improved, smart engineers in a handful of places around the world realized that it was becoming possible to make a good electric car. One of the first companies to try it was founded in California in 2003, the brainchild of Martin Eberhard and Marc Tarpenning, who named their company after one of the early geniuses of the electrical age: Nikola Tesla. Needing cash, they took on an investor who would, soon enough, take control of the company. He was a recently minted technology millionaire named Elon Musk.

Mr. Musk's name has since become synonymous with the effort to electrify transportation and end the era of fossil fuels, not to mention his entirely separate push to build spacecraft that can put astronauts on Mars. He is without a doubt the most fascinating—and, sometimes, the most annoying—technologist of the modern age. Mr. Musk is something of a cross between Thomas Edison and P. T. Barnum. Tesla is a high-flier in the stock market—worth ten times more than, for comparison, the Ford Motor Company—that has managed to interest people all over the world in electric cars. Exploiting advances first made by other companies in producing lithium batteries at scale, Mr. Musk has managed to take the century-old dream of an electric car and turn it into one of the sexiest ideas of our time. We use that word deliberately; here are the four model names of the cars that Tesla has brought to market: Model S, Model 3, Model X, and Model Y . . .

S 3 X Y.

Yes, that "3" was originally supposed to be an "E," but the Ford Motor Company blocked Tesla from using it, claiming "Model E" sounded too much like Model T.

When Tesla first laid out its ambitious plans, Mr. Musk explicitly adopted technological learning as his operating principle. "Almost any new technology initially has high unit cost before it can be optimized and this is no less true for electric cars," Mr. Musk wrote in 2006. "The strategy of Tesla is to enter at the high end of the market, where customers are prepared to pay a premium, and then drive down market as fast as possible to higher unit volume and lower prices with each successive model." If there's a pithier summary of how the learning curve works, we have not seen it.

Tesla launched with a low-volume, high-priced, handmade roadster as a way of raising early capital and teaching its employees the art of designing and manufacturing cars. Then followed the more refined, but still pricey, Tesla Models S and X. In 2017 the company finally came out with its mass-market car, the Model 3, which immediately seized the bulk of the electric-car market in the United States. Along the way, Tesla solved the range-anxiety problem for its customers, not only by installing better batteries but by building its own far-flung network of rapid charging stations. Those stations were long open only to Tesla owners, but they may be open to everyone by the time you read this.

Mr. Musk's company has had growing pains, many of them self-inflicted, but he has nevertheless scared the world's major car companies into launching electric models of their own. A wave of large car companies has begun investing heavily in electric vehicles—including Volkswagen, trying to chart a new course after its diesel emissions cheating scandal. In early 2020, General Motors, the storied American company, announced an "aspiration" to sell only electric cars after 2035. Other companies have also committed to rapid electrification.

Once again, the magic cycle has begun: as large battery packs scale up, they have jumped onto the technological learning curve, and prices are falling fast. We believe there is potential for a massive switch to electric cars over the next two decades. Thanks to that once-obscure tax incentive, it has already happened in Norway, where electric cars now have more than half the new-car market, and they are gaining share fast in some other countries. In the overall global car market, electric cars have just surpassed 7 percent of new-car sales.

As electric cars move down the learning curve and battery costs fall, that will open important new opportunities to electrify much of heavy transportation as well. Experiments are already underway in some countries with electric trucks, including long-haul trucks that might draw their power from overhead lines strung along the freeways, in the same way that overhead wires are now used to power trains. Tesla plans to make electric trucks running on batteries, with no resort to overhead power. And electric buses are already here, with many American cities starting to purchase them; one Chinese city, Shenzhen, has converted entirely to electrified buses, seventeen thousand of them, as well as electrified taxis. Didi, the Chinese version of Uber, is about to go all-electric in Shenzhen. The city has done all this in less than a decade.

The United States government, and several European governments beyond Norway, now offer tax breaks or other financial subsidies for the buyers of electric cars. This policy of using public money to subsidize private transportation is hard for some people to swallow, including many of our friends in the environmental movement. We think it can be justified by immediate environmental benefits, including cleaner city air and less global-warming pollution, and on grounds of national security, reducing oil imports from potentially hostile nations. But the biggest reason to subsidize electric cars and to accelerate their adoption is to keep pushing batteries down the learning curve. We have already mentioned the way that falling battery costs are likely to benefit the electric grid: companies are putting up big banks of batteries that can store renewable energy for use at times of high demand, eliminating the need for certain kinds of gas-fired power plants. The potential market for batteries in cars dwarfs the market for grid batteries, so continued progress on the electric grid has come to depend, oddly, on how fast the electric car market scales. The combination of an electrified car fleet and a power grid running on renewable energy offers an enormous opportunity to cut greenhouse pollution and clean up dirty air.

It can take a while to charge an electric car—roughly half an hour with a fast charger that you might find along an interstate highway, or four to ten hours, depending on the vehicle battery size, with a "Level 2" charger that you might install in your garage at home, using a circuit like that for a clothes dryer. Fast-charging batteries are on the horizon, but for now, slow charging means we need a lot of chargers—potentially far more than gas stations, which can fill up a car in five minutes. In a suburban setting, this is a fairly easy matter to solve: hook up a charger to the same type of circuit a clothes dryer would use and charge the car in the garage. This is obviously more complicated in an apartment building, never mind a skyscraper. To make things really complex, imagine a city bus depot with a couple hundred hungry buses parked overnight. That requires some serious copper wire.

No doubt this is a real project, and it will not be accomplished overnight. Some people fret that it cannot be accomplished at all; with so few cars on the road, chargers can be unprofitable to operate—but we need many more chargers to get people to buy the cars with confidence. Chris Nelder, an electric car expert and host of the Energy Transition Show, one of our favorite podcasts, does not buy that argument. "A lot of people say electric vehicles have a chicken-and-egg problem, where people won't buy EVs until there are more charging stations, and nobody will build more charging stations until there are more EVs," Mr. Nelder said in an interview. "But I think that's wrong. We have a chicken-and-*waffles* problem. The vehicles are definitely coming, so we should stop waffling and build some charging infrastructure!" He has a point: the car companies have committed to putting a huge new wave of electric models into showrooms.

But the conversion to electric vehicles is well within the capability of developed countries. And the payoff would be vast, in reduced cost, pollution, and dependence on oil. Considering the total running costs over the lifetime of a car, electric cars are already cheaper than gasoline cars, in part because electricity costs the equivalent of about $1.20 per gallon. However, the up-front cost still deters a lot of people. If the cost decline of batteries continues apace, electric cars are expected to be cheaper to buy than gasoline cars sometime before 2025, opening the possibility of a rapid transition.

Pulling the Levers

The kids refused to give up. Month after month for an entire school year, they showed up at meetings of the Phoenix Union High School District board. Their voices quavering just a bit, they stood up every month to make the same request: electric buses, please.

For Monica Aceves, captain of the track team at South Mountain High School, in a working-class neighborhood of Phoenix, the lobbying was personal. She had watched a friend on the team keel over with an asthma attack and spend a week hospitalized. She had begun to learn about the dirty air that especially afflicts neighborhoods like hers, due to emissions from industrial plants and from dirty diesel vehicles, including the very buses children rode to school. "I like to stand up for my community because they deserve a better environment, a better place to live in," she told the *Arizona Republic* newspaper.

The students' persistence paid off in 2019, when the school board agreed to buy one electric bus—and to take steps toward a broader switch down the road. Cheers erupted after the vote.

A similar situation played out in Montgomery County, Maryland, a suburban county outside of Washington. A group of local citizens, including students, began marching in the streets and sending petitions urging local agencies, including the school board and the county commission, to live up to their own rhetoric about climate change. The school board looked at the potential to switch to electric buses and recoiled at the initial cost, which can be triple the cost of diesel buses. But the electric buses are far cheaper to run and maintain, which meant potential savings over the long term. And the students demanding change did not let up.

"The truth is that we need to act as if our house is on fire, because it is," said one of the leaders of the protests, Rosa Clemans-Cope, age fourteen. She and her older sister, Eleanor, led a group of activists that blocked traffic, picketed the county executive's house, and more—getting themselves arrested in the process. "When young people see no other option for their futures than to take such measures, then we know we have a problem," said Ms. Clemans-Cope, who goes by the nickname Rosie.

A clever company in Massachusetts, Highland Electric Transportation, finally solved the county's problem with an unusual proposition:

it would buy the buses and lease them to the school board, then make money in part by tapping into the bus batteries at times when the electric grid was stressed, selling power back into the grid. The financial package was attractive enough that Montgomery County ordered three hundred buses and committed to replacing all diesel buses eventually, the largest such commitment we know of by a school board.

When people ask us what they can do about a problem as large as climate change, we always think about these stories. Every parent in America has a stake in the issue. Kids are riding dirty buses to school; they are breathing diesel fumes, which can trigger asthma; and those buses are adding to the global-warming pollution that is scorching the planet on which the children are going to have to live. School boards listen to parents. Why isn't every parent marching down to his or her board meeting and demanding a change?

While it may be true that cars and trucks and school buses are manufactured by huge companies and designed to meet national regulations, the practical details of how fast we can electrify our transportation system are under state and local influence. Every day, in every community in America, decisions are being made to perpetuate our dependence on fossil fuels. City governments order new gasoline cars to add to their fleets. Counties order diesel garbage trucks, even though electric models are becoming available. School managers order new diesel buses, barely considering the alternatives.

More than a dozen American cities—including large ones like Boston, Chicago, Houston, Miami, Philadelphia, Phoenix, Seattle, and Washington—have committed themselves to bold climate action, in line with the goals of the Paris Agreement. These cities and others need to use their clout to help solve the practical problems that are preventing rapid electrification. They have an influence on how many chargers will get built, and where. They can make city land available for charging spots. State governments have a huge role, too: they often control fleets of thousands of vehicles, and can choose to electrify them. They can push their electric utilities to build charging stations for cars, or to cut deals with private businesses to build them. Cities and counties regulate taxicabs and ride-hailing services, and can push them to electrify.

While cars are regulated at the national level in most countries, the situation in the United States is unusual. Through a quirk of history, California plays an outsized role in national car policy. The basic reason is that greater Los Angeles has 19 million people driving 15 million cars and trucks, in a region encircled by mountains that tend to hold air pollution in place. By necessity, California started setting its own air pollution standards in the 1960s, under a law signed by then governor Ronald Reagan. When the federal government got into the act in the 1970s, a special provision of the Clean Air Act left the state's authority intact. Other U.S. states are allowed to choose between the federal standard and the California rules, and fifteen states have chosen California. States have good reasons to follow California's lead. If they adopt a policy developed in California called a zero-emission vehicle mandate, they can force automakers to put an extensive lineup of electric cars onto their local markets. In states that fail to do this, citizens have many fewer models to choose from. California is also leading a coalition of states working toward faster electrification of buses and heavy trucks.

Unfortunately, a majority of American states still do not follow California car standards. Adoption of the California standards needs to become a political target in every state. Some governors have the power to do it on their own, but in other cases, legislation may be required. The longer the list of states grows, the more leverage California will have to push the car companies to move faster on the automotive transition. Remember when we mentioned the secret levers that control much of what happens in the economy? These efficiency standards for cars are one of those levers, and it needs to be pulled hard in the direction of cleaning up our vehicles.

Governments have one more arrow in their quiver that we have not mentioned, and it is potentially the biggest arrow of all. Nearly twenty of the world's governments, from Norway to India, have already set a goal that no gasoline cars will be sold in their countries after 2030 or 2035. So far, these are mostly just goals, but as we write these words, countries are considering how to convert them into legally enforceable bans. The California governor, Gavin Newsom, has declared that he wants no more gasoline cars sold in his state as of 2035, though it is unclear how such a policy might be enforced unless the federal government adopted a ban for the whole country. Massachusetts has announced a similar goal, and the new governor of New York, Kathy Hochul, just signed a law setting such a goal for that state.

Even if only a handful of countries adopt firm bans, that will be a powerful market signal that the internal combustion engine is on the way out, and will encourage companies to work harder on electric cars. The more countries that adopt such bans, the more powerful the market signal will be, and the easier other countries will find it to follow the leaders. The United States, the world's most car-dependent country, ought to become one of the leaders, though the politics of that will certainly be tricky. The new Biden administration has not committed to such a ban, but it has announced a target for the United States: that half of new cars sold will be electric by 2030.

Cities can also help alter the psychology around gasoline cars. The mayor of Paris has already declared that no internal-combustion vehicle, no matter how new it is, will be allowed to enter the city center after 2030. That, too, is just a stated goal so far—but we think it ought to become a legally enforceable ban. And we think the same ban ought to be adopted in all the world's major cities. In our vision, if you own a gasoline car as of 2030, you will not be able to drive it into New York, San Francisco, New Delhi, Beijing, London, or any of hundreds of other major cities. In fairness, drivers need to be given years of lead time before such bans take effect. But, once the future bans are announced, anybody living near such a city will think hard about buying anything other than an electric car. This is exactly the kind of jolt the market needs to push electric cars further down the learning curve.

If you have the money, we suggest not waiting for such bans to take effect. Become an early adopter and switch to an electric car now. While we have argued in this book that green citizenship is more important than green consumerism, cars are one area where being a green consumer can really help the larger cause. If you manage to get your hands on a Tesla or a Leaf or a Bolt, show it to your neighbors. Invite them to drive it. Electric cars are so superior that people who try them often become instant converts. We see a real possibility for social contagion as electric cars become more common.

Beyond buying a green car, how can ordinary citizens help to electrify transportation? We have already mentioned the possibility of lobbying school boards to adopt electric buses. We see the potential for citizen coalitions pushing in every school district in the country. If you can't find one in your area, start one—perhaps you can seek out like-minded people with a posting on Nextdoor, on a Facebook neighborhood group, or elsewhere. Expect to be told no when you first approach a school district. Because electric buses are so much more expensive up front than diesel buses, they start to make sense only when school districts look at the lifetime cost of ownership. Districts may need to start with a small order, like the one in Phoenix, with plans to go bigger over time. They can also look into leasing the buses from financing companies as a way to avoid the up-front costs, as Mongtomery County in Maryland did.

In a similar way, we think citizens should approach their city councils and state legislators to ask what is being done to encourage the switch to electric cars. Is your town working with other agencies to add charging stations on public property? Is it converting its own fleet to electricity? What about the garbage trucks? Is the state government offering tax credits to get the cars moving faster? Has it adopted California car rules, to get more models into the dealerships? These are all political targets that need to be tackled now.

Some national organizations might serve as sources of information for citizens who have decided to work on these problems. An

CHAPTER 5

An Urban Planet

As that fateful January day dawned in central Stockholm, politicians and journalists huddled in knots near some of the bridges carrying traffic into the city center. They pulled their coats and parkas tight, shivering in the cold wind blowing off the Baltic Sea. Along several of the bridges that morning, protesters were braving the chill, too, carrying placards.

For many weeks, the city had been in an uproar. The small Green Party, which held the balance of power in the Swedish parliament, had essentially strong-armed the government into trying out one of its pet projects. On that morning—January 3, 2006—and for months thereafter, anybody who needed to drive into or out of central Stockholm was going to be subjected to a social experiment. The Greens wanted to find out if a "congestion charge"—a fee for taking a car into the crowded central city—could reduce traffic and clean up the air.

The idea had been discussed around the world for decades. Basic economic theory suggests that any commodity priced at zero will be overused—and you can think of the traffic capacity of the streets as a commodity, of sorts. If you really wanted to get a handle on congestion, transport economists argued, you needed some kind of price signal for private cars. But, fearing political backlash from drivers, few cities had been bold enough to try it.

Jonas Eliasson, a young transportation economist who had been hired to work on the Stockholm test, was not sure what to expect. Like everyone involved, he had been battered for weeks by conflicting opinions from friends and neighbors. He had even heard from his aunt, Gunnel Hahne, who lived nearly two hundred miles west of Stockholm—she favored the new tolls. "She thought it was the right thing to do," he recalled in an interview for this book.

Running the plans through computer models of the Stockholm traffic pattern, Mr. Eliasson and his colleagues kept getting an answer they did not quite believe. The models said the charge Stockholm was about to impose—a little over $2 in each direction at peak times—would suppress traffic by as much as 20 to 25 percent. As the test dawned that morning, he huddled in a City Hall annex with other experts, watching a display of traffic numbers as cars crossed the new electronic cordon around the city. With the public so hostile and the press coverage so negative, the goal was to head-off misinformation by providing journalists with hard facts in real time.

What actually happened that January day in Stockholm has become the stuff of legend, not just in Sweden but to urbanists around the world.

By midday, it was clear the decline of traffic had far exceeded anyone's expectations. Without the usual backups, the streets looked nearly empty, even though they were still carrying quite a bit of traffic. The numerical result appeared the next morning in a front-page headline in the Metro newspaper: "Every Fourth Car Disappeared." The computer calculations had been dead on. Picking up the newspaper, "I gaped in astonishment," Eliasson wrote later. "We had all the figures—journey times, traffic flows—that said the same thing, but the pictures made the point so much more clearly."

The biggest surprise, perhaps, was how much better life became for people who *did* decide to pay the toll and drive into the city.

Without the cascading congestion typical of city centers, journey times plummeted—some forty-five-minute drives fell to twenty-five minutes. Even after people got used to the idea that they would have to pay, traffic was persistently down about 22 percent over the course of the seven-month test.

Opinion polls registered the shock: people could not believe how much nicer central Stockholm suddenly seemed. Traffic was less frenzied, bicycling safer, the air cleaner—and restaurants and shops were still doing a brisk business. The public opposition to congestion charging collapsed. True to its word, the Swedish government stopped the test and put a question on the ballot: Did the people want a permanent congestion charge for Stockholm?

When the votes were counted, the congestion charge had lost in the suburbs around Stockholm, to no one's surprise. But the voters whose quality of life was directly at stake, those living in the municipality of Stockholm, voted in favor of it, 53 percent to 47 percent. Today the congestion charge—now about $5 in each direction at peak times—is a permanent feature of life in Stockholm and enjoys broad support. Reducing greenhouse gases from transport was not a central goal of the project, but the government did track them, finding they fell more than 10 percent in the city and a few percentage points across the larger Stockholm region. Airborne particles that can lodge in people's lungs fell about the same amount. Researchers later confirmed a sharp drop in hospitalizations for asthma once the congestion charge was in place, saving money in the Swedish health-care system—and likely saving lives.

Stockholm is not the only European city that has had a positive experience with congestion charging. London launched such a system three years ahead of Stockholm, inspired by the long-running success of a congestion charge in Singapore. And, at long last, the idea of congestion charging is spreading to the United States. New York City has imposed a congestion charge on taxicabs and other for-hire vehicles operating in lower Manhattan and is moving forward with plans to impose a stiff charge on private cars. The idea is under discussion in Los Angeles, Seattle, San Francisco, and Chicago.

Congestion charging is only one aspect of a broader set of urban reforms that are being adopted all over the world. Paris, through long and diligent efforts that have included shutting down major roads and turning them into parks, has managed to cut car trips in the city center nearly in half. In American cities from coast to coast, city streets are going on a diet: a traffic lane, or sometimes two, is being subtracted to make way for safer bicycle lanes, wider sidewalks and trees. Along these "complete streets," business in the stores often goes up, for the simple reason that the streets are more pleasant for pedestrians.

In the previous chapter, we talked about the need to get rid of fuel-burning cars and convert entirely to running on electricity. But saving the climate demands more: instead of simply switching one type of car for another, we need to rethink our relationship with cars altogether. This is starting to happen: all over the world, the long tyranny of the automobile on our city streets is finally being challenged. The victories are too tentative and too few, but they are accelerating.

We believe anyone dedicated to cutting greenhouse gases and salvaging the climate must embrace this broader urban agenda, whose goal is nothing less than to make our cities more livable. Even if carbon emissions were not an issue, these changes would make sense, but in fact the difference in emissions between a car-dependent city and a walkable city can be remarkable. In Houston, one of the most car-dependent cities in the world's most car-dependent large country, the average person travels almost thirty-seven miles per day by automobile. Contrast that with Chicago, a city with a dense urban core and a workable mass-transit system extending into its suburbs: the comparable figure is just over twenty miles per person per day, a reduction of 46 percent.

Fixing cities takes political courage and public support, and it also takes time. The many changes that drove down traffic in central Paris have taken decades. For congestion pricing in New York, just getting a law in place took more than a decade of political battles. Los Angeles is in its third decade of trying to turn an agglomeration of freeways and suburbs into a livable city with a workable transit system, and that project is incomplete. As the political will to tackle climate change and its related problems rises, the new urbanism is one beacon of hope—but some big questions about it loom. Can we speed the transformation of cities in the developed world enough to make a real difference in emissions before 2050? And in the chaotic urbanization underway in the developing world, can humanity avoid repeating the mistakes of the past?

What Went Wrong?

For the fortunate people of the world, the 3 billion or so who live in the most advanced countries or in the more prosperous cities of the developing world, life is better by many measures than it has been at any point in human history. Once-fatal diseases can be cured. Life expectancy has soared. The likelihood of dying in childbirth has plummeted. Clean water, adequate food, indoor plumbing—these amenities are taken for granted. Technology has put miracles in the palms of our hands.

But in the decades of the twentieth century when all these improvements were occurring, citizens of the Western world nearly forgot something they used to be good at. They stopped building cities that people enjoyed. If we hail from a typical American city, we sense that something has gone wrong when we visit older cities abroad that are vibrant, walkable, and alive. If you go to Amsterdam, with its parks and cafés and street life, and then fly home to, say, Atlanta, you may find yourself walking through the densest part of downtown staring at the blank walls of tall buildings and vast parking lots, wondering exactly what happened.

The short answer is the automobile. Many of those blank buildings in Atlanta were designed with the idea that people would drive their cars into underground garages and take an elevator to the office without ever emerging into the light of day. Many people do exactly that—though they often get to the office exasperated after spending an hour or longer driving from the suburbs down hopelessly jammed Interstates 75 and 85. From a distance, downtown Atlanta looks like a real city—and in fact, some of the older, urbane neighborhoods are delightful—but when you get close, you see that too much of the central city is dead and boring. That did not just happen; it was allowed to happen.

Even if most of us can sense the problems, it takes practiced eyes to know exactly what modern cities are doing wrong. Streets are the arteries of a city, and the typical American city has given that highly valuable real estate over to cars almost entirely. In some downtowns, streets and parking lots take up half the land. This is a radical change from the way streets used to be shared by pedestrians, horses, delivery wagons, bicycles, streetcars, and even trains. After World War II, all other forms of conveyance were evicted. As cars took over, the long decline of mass transit began. Many American cities had better mass transit options a hundred years ago than they do today. You will sometimes stumble across the old trolley lines, now disused and rusting. In the second half of the twentieth century, these systems were starved of investment in favor of the new king of the highway, the car.

Today, the supremacy of cars is so accepted in the United States that many people are blind to how skewed the law has become against walking and bicycling. City traffic engineers widen roads in the name of making them safer, but what they are really doing is creating mini-freeways through urban neighborhoods—speeding traffic up, when they should be working to slow it down. City codes typically require building developers to install far too many parking spaces, essentially socializing the cost of parking by forcing shoppers or apartment renters to subsidize it. The same codes too often permit the builders to put up those blank walls on urban streets. Public policy should deliver streets, and streetscapes, that are amenable to people who walk, ride a bike, take transit, or just want to sit outside on a sunny day. The cities that do the best job of urban planning these days simply reject development plans with blank walls, requiring developers to propose shops at street level, and to include pedestrian- and bike-friendly amenities like small parks or plazas, outdoor seating, and bicycle racks.

Bad as they are, though, the planning atrocities these days can't really hold a candle to that great sin of the twentieth century: the construction of the urban freeways. Freeways are badly misnamed, for in no sense have they really been free. Their costs have been staggering. Yes, the construction of 42,500 miles of American interstate highways in the mid-twentieth century was an epic feat of engineering; it was the largest public works project in history up to that time. For the politicians of that era, it was a leap of imagination to see that it could be done. However, it was also the moment when Americans surrendered completely to the automobile, remaking our urban geography and turning the nation into the oil hog of the world. It was no accident that American commercial passenger-train service, operated for much of its history as a profit-making private business, became dependent on public money at almost exactly the moment that the Interstate Highway System was completed.

Perhaps it was inevitable in the age of the automobile to connect American cities with fast roads, but was it inevitable to pierce the hearts of our cities with those highways? In almost every sizable American city, one or more freeways were routed directly into the downtown. Apartment buildings, schools, and parks had to be ripped out to accommodate the new highways. What ought to have been the choicest real estate in any city—waterfront property on a river or lake or bay—was often surrendered to freeway construction.

As the freeways were planned, local leaders had to make decisions about which neighborhoods to destroy. You can probably guess that it was not the wealthy white neighborhoods that wound up on the chopping block. The hearts of many historic Black neighborhoods were bulldozed during the freeway binge, and the racism animating these decisions was overt. In his book about segregation in America, *The Color of Law*, Richard Rothstein cited the recollections of Alfred Johnson, a lobbyist for the American Association of State Highway Officials when Congress passed the 1956 law that created the interstate system. According to Johnson, "some city officials expressed the view in the mid-1950s that the urban interstates would give them a good opportunity to get rid of the local 'niggertown.'"

This deliberate destruction of Black neighborhoods during the freeway boom is not just some long-ago historical injustice. To this day, in the remaining neighborhoods on either side of urban freeways, poor Black and Hispanic children breathe automobile exhaust all day long and suffer high rates of asthma and other respiratory problems. In this sense and in others, the fossil-fuel economy has been a continuing assault on the bodies of Black and brown people.

The urban freeways had other pernicious effects. They became the enabling mechanism for white flight from America's cities. The huge tracts of suburbia that grew up after 1950 were made possible by the freeways, and they became the bedroom communities of the postwar middle class. As white people (and, to be sure, a fair number of middle-class Black people) fled to suburbia, their tax money went with them. Clayton Nall, a researcher at the University of California, Santa Barbara, has made the case that this geographic sorting into majority-Black cities surrounded by majority-white suburbs is one of the roots of our modern political polarization, with those suburbs becoming the base for an ascendant conservative movement that disdains the cities and has been willing to abandon them. Only recently has this stark racial divide between city and suburb started to grow more complicated, with the political allegiances of some inner suburbs apparently shifting as they become more ethnically diverse.

In short, many of the most acute American problems are all connected: racism and racial anxiety contributed to suburbanization, which led to the extreme dependence on cars that is now our major source of greenhouse emissions. And yet the story of the freeways is not entirely one of woe. As early as the late 1950s, it began to turn into something else entirely: a tale of democratic empowerment, of citizens reclaiming the destiny of their cities from highway engineers. In this movement from the twentieth century, we see a model for citizens of the twenty-first.

The first big victory against freeway construction came in San Francisco in the late 1950s, when a citizen uprising stopped plans to put freeways through the beloved Golden Gate Park. In the 1960s, the great urbanist Jane Jacobs led a fight to block the New York City power broker, Robert Moses, from plowing down big chunks of two neighborhoods, SoHo and Little Italy, to install a freeway. Across the country, freeway opponents marched, they badgered politicians, and sometimes they blocked bulldozers with their bodies. As the anti-freeway movement gathered force, some projects were shut down after construction had already started. To this day, you can go visit a huge hole in Pasadena, California, where an extension of Interstate 710 was supposed to go. In other parts of the country, freeway spurs sometimes stop in the middle of nowhere, monuments to the incomplete visions of the highway engineers.

Blasting through the hearts of cities with freeways was, of course, anathema to the walkable, human-scale streetscapes that make cities work. Starting in the middle of the twentieth century, planners and urbanists heavily influenced by the work of Ms. Jacobs began to understand how much damage the automobile was doing, not just from freeway construction but through its tyrannical seizure of virtually every city street. Sidewalks were being narrowed to make more lanes for cars; the very term "parking" had come into existence because delightful street-side strips of parkland were commandeered in Washington, D.C., as places to stash cars. And the pollution from car exhaust enshrouded cities in an acrid haze called smog that made breathing difficult and sent children to hospitals.

Those pioneering twentieth-century urbanists began to propose solutions: a reversion to human scale, to walkable and bikeable and lively streetscapes. But traffic engineers were in charge, drivers were a powerful political constituency, and reform was halting. Only now, half a century later, are we starting to see the best ideas being put into effect on a large scale. In North America, two Canadian cities—Vancouver and Toronto—have been among the leaders, creating streetscapes that work not just for cars but for people on foot, bike or scooter. People who visit the Netherlands or Denmark imagine that they have been bicycle-friendly countries since the bike was invented, but it is not true. Their biking cultures, and the extensive facilities that support them, were built up through hard work over the latter decades of the twentieth century. And we do mean extensive: Utrecht, a city in the Netherlands, has a bicycle parking garage with room for 12,500 bikes!

We see this kind of intelligent city-making—a program that we will call city repair—as one of the most important answers to the climate crisis. And it is not just an answer for the rich cities of the developed world. Some of the most exciting developments are occurring in the developing countries that have been urbanizing at a breakneck pace.

The Fast Bus

It was 8:05 on a January morning at a bus station in Kimara Mwisho, a suburban district to the west of the sprawling city of Dar es Salaam, the largest urban agglomeration in Tanzania. The place teemed with riders. People heading toward town rushed to buy tickets, paying the equivalent of 28 cents for adults and 8 cents for students. They jostled for a place in the boarding queue, sometimes forced to wait as newly arriving buses filled up in moments. A casual observer might have been reminded of the overcrowded, inadequate bus systems typical in developing countries.

But this bus station, and the larger system of which it is a part, have been a godsend to the people of Dar es Salaam. It is a special kind of bus system, a type that is playing a role in unsnarling commuting times and combating traffic all over the world.

Emmanuel Herman, a veteran photojournalist and resident of the Kimara district, lives about seven miles outside the city center. He used to drive to work, and covering that distance on Dar es Salaam's jammed street grid would take him as much as two hours. Now he no longer bothers with a car: the new bus system can get him there in forty minutes, and if he times things right and catches an express bus, he can make it in half an hour. Not only has his car use declined, but his quality of life has improved. "Back in those days, you didn't have time to spend with the family, but now you arrive home early before the kids go to bed," Mr. Herman said in an interview for this book. Charles Mapuli, who works at a vocational training institute in Dar es Salaam, enjoyed a similar shift. He estimated the bus network had chopped his daily commute time from three hours to eighty minutes.

What is the trick that makes the new system so valuable to people? Over much of the distance the buses cover, they travel in their own dedicated lanes, from which cars are barred. The system is still a work in progress, but when complete, the buses will be able to transmit a code to traffic lights, causing them to turn green as the bus approaches. Even in its early form, the system, which opened in 2016, has been so popular that the authorities are racing to expand it, with plans to put hundreds more buses onto the street. "There were serious transport challenges before the project, such as heavy traffic and wastage of time, which were costing our economy," said Ronald Rwakatare, chief executive of the Dar es Salaam Rapid Bus Transit Agency, the agency that oversees the system. "It has completely revolutionized the public transport."

Dar es Salaam does not have a subway, and Tanzania does not have the billions of dollars it would take to build one. But the city is one of the fastest-growing in the world, with the population of the metropolitan area expected to double over the next decade, to 10 million—making Dar es Salaam one of the world's "megacities." The type of bus system the city is building is called Bus Rapid Transit, or BRT, and when all the pieces are in place, it should achieve speeds similar to those of a subway for less than a tenth of the cost. The government even has plans to run the buses on electricity, after a dam is completed to provide clean power.

The alternative to great public transportation options, for cities like Dar, is to drown in rising automobile traffic and the pollution it brings. That has already happened in New Delhi, in Mexico City, in the booming industrial cities of China—in too many places. Finding an affordable and practical way to move citizens around is critical to improving the quality of life in these cities—and, as we explained earlier, to holding down the emissions that cause global warming. But taking these steps will be politically difficult. The benefit of cutting emissions may feel a bit theoretical if you are being asked to consent to something that will ding you in the wallet, like a new fee for driving your car into the middle of town. But if those same measures are likely to benefit you in a direct and immediate way, such as through improved public transit and walkable streets, we think it becomes easier to support them. Indeed, forward-thinking people are starting to demand them.

City Repair

The New York City bureaucrats were nervous. Like their predecessors in Stockholm, they were about to try a social experiment on a reluctant city. They had convinced the New York mayor, Michael Bloomberg, to shut down a big segment of Broadway to turn Times Square, where Broadway met two other streets at confusing angles that produced a near-permanent traffic jam, into a pedestrian plaza.

The city had ordered chairs to be put into the street, but they were late arriving, producing a bit of a panic as the big day neared. A rushed search led the city planners to Pintchik's hardware store in Brooklyn, where they snapped up 376 beach chairs in vivid colors for $10.74 apiece.

When the day arrived to close the street, nobody had any real idea what would happen. The tabloids were predicting traffic doom. Cabbies were furious at losing one of their most important routes through one of the world's busiest cities.

Janette Sadik-Khan, the transportation commissioner who'd convinced Mayor Bloomberg to try the plan, gave the order to go forward. Orange traffic barrels rolled into place, closing off Broadway. Into the middle of the street went those 376 beach chairs. Contrary to predictions that New Yorkers and visitors would hate the new plaza of black asphalt, they poured into it by the thousands. Within minutes, you could not find a beach chair to save your life. People laughed. They sipped drinks. They snapped photos. Day after day and far into the night, the scene was repeated. Times Square was already famous for its street life, but now that daily circus of performance art blossomed. People suddenly had plenty of room to gather around the buff Naked Cowboy and watch his act. Within a few months, the city had declared the shutdown permanent, closing a stretch of Broadway all the way from Times Square to Herald Square, seven blocks south. Permanent bistro tables and chairs eventually replaced the beach chairs.

Over the coming years, business in the area would boom as Times Square underwent a broader urban makeover that made it one of the world's shopping destinations. Commercial rents would double, then triple. Other cities would copy New York, reclaiming space from cars and handing it over to pedestrians. New York itself would go on to create more than sixty of these plazas across the five boroughs.

Here is the craziest thing, though: With Broadway closed off, traffic through Midtown Manhattan *improved*. How could that be? you ask. It was actually pretty simple: by taking out a street that met other streets at weird angles, the city was able to unsnarl confusing intersections, giving both drivers and pedestrians a simpler route and longer green lights to make their way through. Even with less road capacity, traffic closest to Times Square moved 7 percent faster. Injuries plummeted more than 60 percent. Because cars sitting in traffic, idling, waste so much fuel, it is likely that the emissions of cars moving through the area fell, too.

This is just one small, granular example of the efforts going on around the world to reclaim public space from the automobile. The changes in New York, like the ones that have cut Paris traffic in half, have basically come from the top down. But in cities and towns across the world with less enlightened administrations, citizens are sometimes taking matters into their own hands.

The movement is called tactical urbanism. What it means, more or less, is: let's try reclaiming some public space to see if it works. The goal is generally to get permanent changes made by putting on temporary demonstrations. Citizens have used temporary paints to put crosswalks onto streets that need them. They have installed temporary bike lanes using paint and traffic cones. They have turned disused parking lots into pop-up art exhibits or cafés. It does not work every time, but cities have made hundreds of changes based on this new form of protest. We see this movement as a democratization of urban planning, an effort by citizens to reclaim the street not just from the automobile but from the traffic engineers who blindly do the bidding of drivers, at the expense of other street users.

Yet in this great project of city repair, fixing the streets is not enough: we also need to tackle the problem of density. Many cities face a housing shortage that is driving up home prices much more rapidly than incomes are going up. The problem, at least in part, is city zoning rules that have essentially made it impossible for builders to put up enough urban housing to meet the demand.

Unsnarling this tangle of issues is going to be difficult, but cities have begun to try. In numerous areas dominated by single-family homes, add-on apartments—"granny flats"—have recently been legalized. This means you can build an apartment behind your house or over your garage and rent it out. These flats can be used by singles or students, as starter homes for small families, and as a reasonable option for retirees. The other benefit? The rent helps strapped homeowners pay their mortgage. The boldest step cities can take is to eliminate single-family zoning entirely. Minneapolis recently did, with the measure passing the city council twelve to one after a contentious public debate. Wherever houses stand now, it will become legal in Minneapolis for developers to replace them with apartment or condominium buildings of up to three units. That doesn't mean single-family houses have become illegal; nothing will stop builders from putting them up if that is what the market demands. But if the market wants denser housing, homeowners or developers can convert single-family lots to house more people. The transition will take decades, but Minneapolis is likely to become a denser, more lively, and more affordable city. This leap toward greater density can also be instituted at the state level. Two states, California and Oregon, recently passed laws that effectively override single-family zoning across large areas. Both laws give landowners in urbanized zones the right to build up to four apartments on lots that typically host only one house now.

Increasing the density of our cities will help with mobility, indirectly cutting carbon emissions. Trains and buses make more sense, of course, the denser a city is. One of the outlandish failures of recent decades has been a refusal of cities to allow, or encourage, high-density homes and shops in locations right next door to transit stations. You can see this in San Francisco, Los Angeles, Atlanta, and many other places: expensive train lines running past miles of low-slung suburban development. By contrast, some of the towns outside Washington, D.C., have done a good job of building dense housing clusters near train stations, putting thousands of people within easy walking distance of public transit. Another California law that passed in late 2021 will encourage high-density housing development along transit lines, in time correcting some of the failures in San Francisco and Los Angeles.

The biggest density challenge is, of course, the obvious one: the American suburb. What do we do about the thousands of square miles of tract houses that ring American cities, a style of living that is entirely dependent on the automobile? Because it forces people to drive virtually everywhere, this development pattern must be seen as a major aspect of our climate problem. One answer is to densify the suburbs wherever it is politically feasible to do so. Land prices are starting to favor this prospect: new building lots have become so expensive and scarce in many suburbs that quite a few have already altered their zoning requirements to allow townhouses or low-rise apartment buildings.

We are also realistic, however: the suburbanization of America did not happen overnight, and it is not going to be reversed overnight. And there are ways for you to be part of the solution without having to move. If you live where you must have a car, we urge you to buy an electric car. If you live in a big house heated with gas, we want you to tighten up the house and convert to a heat pump, which uses much less energy—and that energy can be supplied from renewable sources. In earlier chapters, we outlined a set of policies—particularly strict building codes—to cut energy use in newly constructed buildings. Suburbs are so wasteful by their very nature that, anywhere single-family houses are still being built, it is imperative that such policies be put into place without delay.

How to Build a City

We have discussed the repair of older cities, but the problem is much bigger than that: new cities are being built all over the world. The human future is an urban future. Today, roughly half the world's population of 7.5 billion people live in cities. By 2050 or so, when we have grown to perhaps 11 billion in number, some 8 billion people will live in cities. In other words, the urbanized part of the world will double in population in the next few decades. Driving the trend is not just population growth, though: we are also in the midst of a tidal wave of people moving from the countryside into cities to improve their job prospects. This happened in rich countries a century ago, and it is now happening all over the world. We are adding a tranche of urban development the size of Paris to our planet every single week.

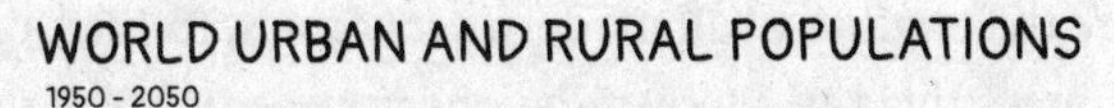

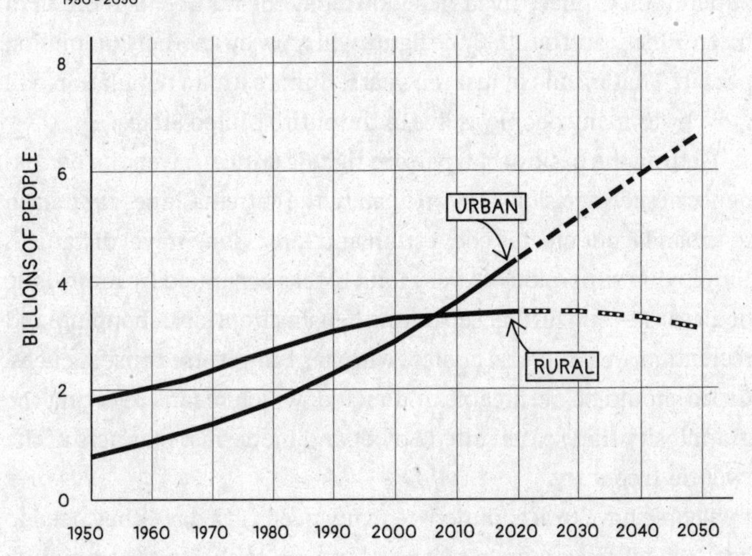

The urbanization of our planet means that a huge construction boom is already underway and destined to get larger. Fortunately, if development is done right, cities will not only produce fewer greenhouse gas emissions, but their residents will get cleaner air, less traffic, less noise, and more options for shopping, entertainment, and education within walking distance. If you are starting from scratch, as many countries are, how do you build cities that achieve those goals?

In the life of a new city, early, pattern-setting moments really matter. Long ago, planners like Daniel Burnham in Chicago and Pierre L'Enfant in Washington established patterns that have affected us ever since, influencing how we get around, how people's work and home lives connect, and so on. Today China is in that kind of formative moment. The country is undergoing the fastest, largest social transformation in human history. More than 500 million Chinese have been lifted out of poverty in the last three decades. This was accomplished in large part by people moving from the countryside to cities, which are the engines of all modern economies. In 1978, only 200 million Chinese lived in cities; today the number is more than half a billion, and by 2030 the figure will grow by another 300 million people. That means in just ten years, China's urban population will grow by as many people as live in the entire United States.

Getting the basic urban pattern right is critical, given the permanence of such decisions. Unfortunately, the current Chinese approach to urban layout creates congestion and forces long travel distances. Employing superblocks—very long blocks separated by multi-lane boulevards—and further separating housing from jobs, shopping, and recreation are structural choices with hard-to-reverse consequences. Cities should be permeable and mixed, which means rejecting the grandiosity that so delighted Soviet and modernist planners of the twentieth century.

People have to accommodate many needs in a day—they need to go to school, to work, to local parks, to stores, to clinics, to community events, to the laundry, to the gym, and so on. Separating all those uses by miles forces people into cars. Lots of cars times lots of travel equals traffic, traffic jams, and excessive emissions.

Today, fewer than one in ten Chinese own a car—yet China is already experiencing massive traffic congestion. In 2010, Beijing earned the dubious honor of enduring the world's largest traffic jam, stretching sixty-one miles, and requiring eleven days to untangle. It is clear that increased automobilization now leads to *decreased* mobility.

The idea of moving people out of cars and getting them to walk or bike is not utopian dreaming, either in China or in the Western world. In Amsterdam, with its famously drizzly weather, a third of trips are made by bike and almost as many by foot, leaving the car share with just 20 percent of the total. Among large cities in North America, Vancouver has been on a drive to make bicycling more practical, with an extensive network of bicycle lanes protected from moving cars.

The recent advent of electric-assist bicycles changes, sometimes radically, the prospects for urban biking. The range people will ride on an electric bike increases from a couple of miles to twice that. Carrying groceries or kids becomes trivial. Hills flatten. And there are no more sweaty clothes at work. China now has more than 300 million electric-assist bicycles, and biking is sharply on the rise, as the engineer-planners there have come to understand that they can use limited street space far more efficiently for bikes than for cars. Biking and walking are not the answer to every transportation problem, but they can be a big part of the answer.

If a city gets the basic street grid right, perhaps no question bears on its success—or its emissions—more than public transit. As Bogotá mayor Enrique Peñalosa says, "an advanced city is not one where even the poor use cars, but rather one where even the rich use public transport"—because it is the fastest and most convenient way to get around. If mass transport is a second-class option—slow, unreliable, inconvenient, dirty—it will be the choice of last resort. Anyone who can do so will switch to a car. And every person who does that reduces the system's revenue, so service declines, spurring more departures, spurring more declines.

The setup we discussed earlier in this chapter, Bus Rapid Transit, was pioneered in a city called Curitiba, in Brazil. Done right, it can be an ideal solution for fast-growing cities of the developing world that cannot yet afford billion-dollar subway projects. As we mentioned, BRT systems can achieve subway speeds and capacities at less than 10 percent of the cost of building a new subway. The trick is to cleverly use surface streets, coupled with ingenious bus configurations, to allow buses to bypass traffic and to avoid the bottlenecks that occur when buses stop on every other block to load passengers.

A high-quality rapid bus system will have station platforms with fare gates so that riders have already paid by the time they board the bus, speeding departures. The buses have a whole wall of doors, like a subway car, so that dozens of people can pile on at once. Guangzhou, in China, recently completed a rapid bus system that moves eight hundred thousand people per day at high speeds, and China has committed to building more than three thousand miles of this type of bus corridor over the coming five years. We are accustomed to new inventions popping up first in wealthy countries and eventually diffusing to the developing world, but BRT is moving in the opposite direction. In the United States, cities like Indianapolis; Pittsburgh; and Richmond, Virginia, have all built versions of Bus Rapid Transit systems. Ridership on the new Richmond system has been double the level that was forecast before it was built. The city has capitalized on that popularity by rezoning land along the bus corridor to allow the construction of as many as seven thousand new apartments, increasing urban density. Like their counterparts in Dar es Salaam, the transit managers in Richmond are now scrambling to expand their system to keep up with the crush of public demand.

If this approach can be made to work in cities as disparate as Richmond, Dar es Salaam, and Guangzhou, it can work anywhere.

Pulling the Levers

The coffee urns came from a nearby restaurant. People rustled up chairs, planter boxes, and old-fashioned street lamps. The dingy facades of empty storefronts got a fresh coat of paint, and artists took up residence in one of the old buildings. For just a couple of days in 2010, a group of people living in a neighborhood southwest of downtown Dallas decided to show City Hall just how you might go about fixing neglected urban spaces. On a street where the city could not be bothered to install a bicycle lane, the group hauled out cans of green spray paint and made a bike lane themselves.

"It was this punk rock, stick-it-to-the-man thing," the leader of the Better Block Project, Jason Roberts, told the *Houston Chronicle* newspaper. "We flouted every law we could." In fact, his group posted placards nearby explaining which of the city's outdated zoning regulations they were deliberately violating. That included a law requiring a $1,000 permit to have flowers in the street, among other stupidities. As the street came to life that weekend, the neighbors waited for the cops to show up and arrest them.

That is not quite what happened. "Hordes of people showed up," reporter Lisa Gray wrote in the *Chronicle*. "They bought flowers and food, and they hung out at Philosophia Cafe, pumping coffee from urns and eating pastries. Bands played. Pedicabs hauled people back and forth. Guys played chess. Kids drew on the street with sidewalk chalk."

Not only did a lot of the neighbors show up, but so did members of the city council. They were not mad. Instead, they had the good sense to ask their constituents for advice about how to make Dallas a better place. Within a year, Mr. Roberts and his team were advising the city on how to bring life back to the dead plaza in front of City Hall itself.

What the neighbors in Dallas had built that weekend was basically a "complete street"—a street that cars can pass through, yes, but which has been narrowed to accommodate activities other than the automobile. Thanks to that *Houston Chronicle* story, the episode in Dallas would achieve fame as a prototypical example of "tactical urbanism," the effort to force reluctant city bureaucracies to accept changes by presenting them as a fait accompli.

The forces that have shaped modern American cities, driving street life out of them and turning them over to the tyranny of the automobile, can seem impersonal and unapproachable. But these bad decisions were made by people. They can be undone. Citizen activists are proving it, all over the world. A tactical urbanist group in Portland, Oregon, actually named itself "City Repair" and has managed to revive a slew of intersections there. The idea has spread to towns across the country.

We think this urbanist agenda dovetails with the climate agenda. People will ride bicycles to work or the grocery if you give them safe corridors on which to do so, leading to fewer car trips and lower emissions. They will take transit to work if it is efficient and affordable, leading, again, to fewer car trips. But these changes require politicians to have courage, to stand up to the drivers or shortsighted business interests that may oppose change.

This is why people who care about the climate need to raise their voices in support of complete streets, in support of bike lanes, in support of closing some streets to cars entirely. If a city's bureaucracy won't move, we need to haul out paint cans and put bike lanes and crosswalks in the streets ourselves. We need to go down to City Hall and make clear that we support wiping single-family zoning off the books, so that a gradual increase in the density of neighborhoods can occur. We need to speak up in favor of congestion charges in traffic-choked cities, with 100 percent of the money raised from these charges devoted to improvements in public transit.

We need to oppose freeway expansions, anywhere and everywhere they are proposed. Expanding the freeways to try to solve traffic is a fool's errand; the wider roads simply draw more cars, so that traffic deteriorates back to the starting level. This problem, called induced demand, was first recognized in the 1930s, yet traffic engineers never seem to learn. The absurd pursuit of ever-more freeway capacity has reached its apotheosis in Houston, where one segment of the Katy Freeway stretches twenty-six lanes across, counting the access roads. A fight is underway in that city over state plans to expand another highway there, Interstate 45 north of town. The expansion will draw more traffic, put more pollutants into the air, harm the lungs of more young people in the twenty-six schools near the freeway, and encourage more car dependency. Bakeya Nelson, a Houston community activist who led a coalition that opposed the project, argues for a national moratorium on urban freeway expansion while the effects on nearby communities of color are assessed.

Not only do we agree with her, but we think many urban freeways actually need to be torn down. Climate advocates need to speak up in favor of doing so. It has already happened in quite a few cities; Boston moved its Central Artery underground, and after an earthquake damaged the Embarcadero Freeway in San Francisco, the city knocked it down. The effects of freeway removals have almost always been positive, with urban development and street life returning to neighborhoods that had been ruined. A group called the Congress for a New Urbanism publishes an annual "Freeways Without Futures" report, a target list of highway segments that ought to be removed. The latest version has fifteen freeways on the hit list.

That's a good start, but we need to do more. The historical injustices caused by freeway construction projects must be corrected. In 2021, at long last, Congress put up $15 billion for projects that will reconnect historically Black neighborhoods severed by the freeways. In Atlanta, for example, planning is underway to cover a twelve-lane expanse of freeways with a huge platform—a "freeway cap"—to hold fourteen acres of parks, streets, and transit facilities. These will be terrific new amenities for the neighborhoods that the construction of Interstates 75 and 85 nearly destroyed.

Appropriately enough, the plan is called the Stitch. And that strikes us as an apt metaphor for the larger program we all need to embrace. We need to stitch America's urban fabric back together. Our climate depends on it.

CHAPTER 6

People, Land, and Food

Forests are the glory of New England. In the spring, the broadleaf trees covering millions of acres shake off the winter snows and send leaves bursting forth: red oaks, white oaks, black oaks, birches, beeches, and maples. At slightly higher and colder elevations, evergreens like spruces, firs, and white pines may be mixed among the deciduous trees. In the autumn, the dying leaves paint blazes of flame red, golden yellow, and burnt orange across the countryside. Tourists flock to New England just to gaze at the autumnal landscape. If you go strolling through one of these forests in high summer, you may imagine it has been there forever. Oak trees 150 feet high will shade you with their broad canopies. You might stop under one of those trees for a picnic, grateful that such spectacular forests have managed to survive the human assault on the environment.

Looks can fool you, though.

Keep strolling through the forests of New England and you will begin to see things that do not quite make sense. In the middle of nowhere, you may come across a wall of precisely stacked stones, seemingly put up by some meticulous human hand—but why would anybody take the time to do that in the middle of a forest? Keep going and you will come across old chimneys standing in the woods. If you clear away fallen leaves and inspect the ground closely around one of these chimneys, you may find splinters or other evidence of the farmhouse that once stood there. Stroll farther and you may see old stone paddocks where pigs and other farm animals were kept.

The truth is that much of the forest land of New England was cleared for farms in the nineteenth century. If you could go back in time, what you would see would be much different—a region of small farms and carefully tended fields, with remnant patches of forest here and there. The landscape would resemble the carefully tended English countryside from which many of the settlers came. "When the European colonists came to America, they saw trees, and they wanted fields and pastures," explained William J. Munger, a Harvard research fellow who studies the region's forests.

In the middle of that century, the farmers of New England were hit by what was, from their point of view, an economic calamity. Once the Erie Canal and later the railroads had opened up the interior of the country, farmers tending New England's stony soils found it hard to compete with crops grown in the rich, black soils of the Midwest. Even as life on the farms grew harder, industrialization was opening a new way of life in New England's towns and cities. One by one, the farms of New England were abandoned. Nature then took its course. Trees colonized the old fields, the farmhouses rotted—and only brickwork and stacked-stone fences were left to testify to a vanished way of life.

Today, the regrown New England forests are not just beautiful tourist attractions; they are one of the most important assets the United States has in battling global warming. The nation's forests offset more than 10 percent of American emissions of carbon dioxide. Careful science has documented that the forests across much of the country are gaining weight as they pull carbon out of the air and sock it away in their tissues as carbohydrates and proteins. The trees are functioning as a sort of sponge for our emissions of carbon dioxide, and this is also true at a global scale.

How human beings use land—how we manage forests or mismanage them, how we run our farms, what we choose to produce and eat from those farms—has profound consequences for the climate. The destruction of forests, wetlands, grasslands, and peatlands is one of the major sources of the carbon emissions causing the crisis. This same wanton destruction is also causing a second crisis, the potential extinction of much of the biological diversity on the planet. And yet, even as these tragedies unfold, the land also offers potential solutions, including the possibility of pulling large amounts of carbon dioxide back out of the air, if we can figure out how to do what the forests of New England are doing—but on a far larger scale.

Depending on how they are managed, forests can tamp down or exacerbate global warming. As much as a quarter of the excess carbon dioxide in the atmosphere today may have come from human beings chopping down trees and disturbing soils and wetlands, beginning centuries ago. But unlike industry, which only puts greenhouse gases into the air, land works the other way, too—it can pull carbon dioxide back out. In scientific terms, the world's land surface is both a *source* and a *sink* for greenhouse gases.

The forests of the world are already sopping up a huge fraction of humanity's emissions. Of all the carbon dioxide we are pouring into the air, just under half of it stays there over the long term. The ocean is taking up slightly less than a quarter of the CO_2, and the land surface slightly more. That work is mostly done by trees, which suck many billions of tons of carbon dioxide from the air, strip off the oxygen, and use the carbon as a building block for their wood, leaves, and roots. Agricultural crops do this, too, combining the carbon with water and elements from the soil to create the sugars, fats, and proteins that make up the food we eat, as well as the feed that goes to farm animals.

Humanity is putting an enormous strain on the capacity of the land to support us as the population grows. In traditional societies, both birth rates and death rates for children tended to be high. In industrializing societies, basic sanitation measures began to cut the childhood death rates in the nineteenth century, and after World War II these measures spread around the world. Child deaths plummeted quickly, but the birth rate was slow to follow. The result was an explosion of the human population, which more than tripled in the twentieth century, from 1.6 billion to 6 billion. By now, the population growth rate has slowed worldwide, to the point that some countries are worried about population decline. This "demographic transition" is not quite complete, however: hotspots of population growth remain, particularly in southern Asia and in the African countries south of the Sahara Desert. The human population today is approaching 8 billion and may exceed 10 billion before it finally peaks.

As the population grew rapidly in the middle decades of the twentieth century, many people worried about the potential for mass starvation. Norman E. Borlaug, an agronomist employed by the Rockefeller Foundation to work on wheat yields in Mexico, developed improved crop varieties that could, with heavy applications of fertilizer, radically increase crop yields for poor farmers in the developing world. The spread of these techniques—a landmark development known as the Green Revolution—staved off mass famine. In the 1970s, these techniques reached China, a country with a tragic history of mass starvation. By establishing food security and freeing farmers for urban factory work, the Green Revolution helped to set the stage for China's rise as an industrial power.

Although it staved off famine, the Green Revolution has had environmental and social costs. Achieving high yields of crops like corn and wheat requires artificial nitrogen fertilizer. Nitrogen is indispensable to building proteins, and is thus an essential component of the tissues of plants and animals; but though nitrogen is abundant in the air in a chemically inert form, plants require a different, chemically active form. In traditional agriculture, this was always in short supply. That age-old agricultural constraint was broken in the early twentieth century with an industrial technique that uses natural gas and a series of chemical reactions to pull nitrogen out of the air, then converting it into a form that plants can use. Without this manufactured fertilizer, we would not be able to produce enough food to supply the prevailing diets of more than 40 percent of the world's population. This nitrogen is placed on farmers' fields in the form of ammonia or one of its derivatives, then gets taken up by plants as one of the fundamental chemical building blocks of proteins. Those, in turn, make their way into human bodies. The scholar Vaclav Smil calls this human addiction to manufactured nitrogen a "profound chemical dependence."

In some places, notably China, the use of nitrogen fertilizer is heavily subsidized, and farmers apply excessive quantities of it to try to guarantee a good harvest. Some of the excess can turn into a gas that enters the atmosphere, becoming one of the causes of global warming. But much of it washes into rivers, where the nitrogen can cause blooms of algae that, when they die and decompose, can suck most of the oxygen from the water, creating "dead zones" where no fish can live. Hundreds of these dead zones now occur at least seasonally at the mouths of rivers around the world, and the problem is growing. This overloading of the landscape with nitrogen, a problem called eutrophication, is not as acute in the United States as in China, but it is still serious. A large dead zone occurs every year at the mouth of the Mississippi River.

The human population boom has not only created huge new requirements for fertilizer—it has also placed immense demands on the world's land. At least a third of the world's forests, along with extensive grasslands and wetlands, have been destroyed for agriculture. This human assault on natural landscapes has been the biggest factor causing plants and animals to go extinct. The geological record suggests that mass extinctions have wiped out most life on Earth five times in the past, including the extinction that killed the dinosaurs. Some experts fear that humans are now precipitating the sixth mass extinction in the planet's history.

To stop that, humanity really needs to pull back—to stop exploiting so much of the land surface and return some of it to nature, allowing forests to regenerate and the natural world to recover. That would also help to pull more carbon dioxide out of the air. But population and economic pressures are pushing in the opposite direction. Food demand is going up, not just because the population continues to grow, but because as countries are getting richer, people are demanding richer diets. Meat consumption is rising worldwide, even though meat is the most environmentally costly food we eat. Among the major varieties of meat, the most damaging by far is beef; producing it can emit up to sixty times the greenhouse gases that are produced from growing the same caloric value as vegetables. A primary reason is that the digestion of cows, unlike that of pigs or chickens, produces large amounts of methane gas. Methane is a powerful greenhouse gas, and although it breaks down rapidly in the atmosphere, the emissions from cattle production are large enough to make a significant contribution to global warming.

Agriculture is not the only source of pressure on the world's supply of land, of course, as has already been discussed. Cities are relatively compact compared to agriculture, but in much of the world they keep expanding outward in a wasteful pattern of suburbanization, creating vast low-density carpets of development that depend entirely on the automobile, chewing up forests and farmland in the process. The recovered forests of New England are under this kind of stress, with parts of them being chopped down a second time for suburban development.

On top of all these pressures comes the climate crisis. Rising heat and water stress are becoming another threat to the health of forests and grasslands. The ability of forests to keep taking up a large fraction of our carbon dioxide emissions is at risk as they suffer drought, fires, and attack from heat-loving insects. We need to produce more food on less land, but erratic weather is working against us. And yet even as the stresses rise, better use of land may be one of the answers to the climate crisis.

One group of researchers has calculated that the carbon dioxide uptake of land in the United States could potentially be doubled over time. That could become a major part of the nation's climate solution. Of course, doing this will cost money. And it will necessitate persuading, requiring, or paying millions of landowners across the country to manage their land differently.

The most immediate opportunity is exactly the one you might guess: planting more trees. The term for chopping down forests is deforestation, and it has an antonym: afforestation. China and some other countries have shown that it can be done on an immense scale. But if you are imagining that this sort of thing might be carried out by a few Boy Scout troops and financed with bake sales, think again. The country already has voluntary tree-planting programs, but their reach is limited. Planting trees is expensive: the seedlings have to be started in nurseries, and expertise is required to put the right trees in places where they will thrive. Afforestation can easily cost hundreds of dollars per acre of land, though it's likely this can be cut somewhat if the United States were to adopt national policies to do so at scale.

One possibility would be to tweak the nation's farm programs to encourage more tree-planting, as well as other beneficial land practices. Farmers get extensive aid from the federal government. Those programs are politically sensitive, but it may be possible to achieve some land-use goals by adding more money to the overall pot and then directing it to the right purposes. One particularly beneficial program is called the Conservation Reserve Program, which pays farmers to set aside pieces of land that aren't very good for farming and let them revert to natural vegetation. The program is currently limited to about 25 million acres. We think Congress should expand the program and raise the amount farmers are paid for conserving their land, with a long-term goal of enrolling 50 million acres.

Another beneficial approach would be to expand existing programs that pay farmers to plant cover crops. These are crops like rye, clover, and sorghum that are not intended for harvest but are instead planted solely to help add organic material to the soil. They are generally planted at times of the year when no cash crop would be growing and left to die back in the field. The decay of these cover crops can add organic material to the soil. Known as humus, this material is basically carbon by another name. Agricultural lands hold immense amounts of carbon in this form, and it can be increased if cover crops are coupled with reduced land tillage. The government already runs a national program to help farmers with the cost of cover cropping, but the payments are too low, and the program needs to become more ambitious.

Wild claims are sometimes made about exactly how much carbon could be stored in the soil if the nation adopted different agricultural practices. But again, money is a major issue: planting cover crops is expensive, just as is planting trees. The farmers get a long-term benefit as their soil grows richer, but they are often operating on thin margins, and may make the calculation that this future benefit is not worth the short-term cost. It's unclear how much carbon could really be captured by the kinds of programs we have outlined, as the United States does not even *measure* carbon in the soil on a national scale. We need a set of national experiments, somewhat equivalent to the clinical trials that are used to test new drugs. Congress ought to authorize several large-scale pilot programs, paying farmers to adopt certain practices as scientists compare those plots to conventional plots. The goal would be to figure out which techniques show the most promise for capturing carbon across the whole landscape, under real-world conditions—and to learn what it really costs to get farmers to adopt them. Until the country launches such a plan, states can work to get a handle on these issues.

From citizens, the country needs to see a national political demand for broader use of restorative land practices. Nobody seems to be against planting trees, but you certainly do not see any major constituency marching in favor of it. To build such a movement, citizens can start working at the local level, in cities as well as in the countryside. Indeed, it is in cities where a bit of political effort might help solve one of the nation's least-understood inequities.

If you stroll through the nicest neighborhoods of America's older cities, you may well be shaded by large, lovely trees. Now go over to the poor side of town: decades of municipal neglect, beginning with the racial redlining of the 1940s, have left low-income neighborhoods with few or no trees. This is not a trivial difference—in hot weather, shady neighborhoods can be as much as fifteen degrees cooler than treeless ones. The disparity is beginning to matter a great deal as the climate heats up. Poor people living without air-conditioning are vulnerable in heat waves; a 1995 heat emergency in Chicago led to the deaths of more than seven hundred people, overwhelmingly people of color living on low incomes in poorly shaded neighborhoods. Urban tree-planting programs can be even more costly than those on farmland, but an expansion is clearly needed. The Environmental Protection Agency calculates that urban trees already absorb about 2 percent of the nation's carbon dioxide emissions; more of them will not only help offset the climate crisis in the long run, but potentially save lives in the short run. An ambitious national grant program for urban forestry would be a wise policy. Some cities like Detroit and Phoenix have already started to tackle the problem, making commitments to "tree equity" that will require spending public money to correct the historical injustice.

Saving the Tropics

On a map of the world, two continuous bands of forest stand out, encircling the land areas of the globe. The immense boreal forest, or taiga, is the northern evergreen forest that runs through Alaska, Canada, Scandinavia, and Russia. But the largest and richest band of forest, biologically, is the tropical forest, encompassing the Amazon of South America as well as rainforests and moist forests of Africa and Indonesia.

Tropical forests, with year-round sunshine and plentiful rain, support the planet's greatest diversity of land organisms. People who stay overnight in a tropical forest zone never forget it. The cacophony is incredible, a dense wall of sound emerging from the woods: the howls of monkeys, the chirps and mating calls of millions of insects, the croaks of frogs, the urgent swooshing of bats' wings, and the low woofs of piglike animals called peccaries. In the daytime, iridescent butterflies and wildly colored birds flit through the cool forest understory.

For decades, saving tropical forests has been one of the most popular environmental causes. Yet it is a battle we have been losing. Some 20 percent of the Amazon is already gone. Entire islands of Indonesia have been denuded, prompting emergency measures to save the orangutans once living in those forests. Deforestation in Africa has lagged behind the other hotspots, but it is growing. The great apes may be our closest living relatives in the animal kingdom, but that has not stopped us from wiping out their homes. They are now at risk of extinction in the wild.

The greatest problem in most of these forests is some combination of logging and agriculture. The rising global demand for commodities that can only be produced in the tropics is a major factor. Expanding markets for chocolate, coffee, and palm oil have led producers to cut down large sections of forest. Perhaps the greatest pressure these days, though, is coming from the global meat industry. Some cattle are grown in the tropics and shipped abroad, but much of the damage is less direct than that: land in the tropics is being used to grow feed, like soybean meal, that is then shipped to other countries to supply industrial-scale meat factories. Remember when we mentioned earlier that China had learned to feed itself in the twentieth century? That is no longer the case: the Chinese now import immense quantities of soy from Brazil to feed a booming meat industry, particularly hogs.

Yet Brazil is a major success story when it comes to tackling deforestation. A former president, Luiz Inácio Lula da Silva, came into office concerned about what the rapid loss of the Amazon was doing to Brazil's international image and—with environmental groups pressing boycotts of Brazilian agricultural products— its business prospects. In 2003, he launched a serious effort to get control of the problem, including prosecutions of farmers who were violating laws meant to protect forests. In a decade, Brazil cut deforestation by 80 percent. Under a new, right-wing government beholden to farming interests, deforestation has crept up in recent years, but it is still well below the highest levels of the early 2000s. Unfortunately, the Amazon is also starting to be hit by unusually intense droughts, raising concern that the increasing temperatures from climate change may be a threat to the survival of the forest.

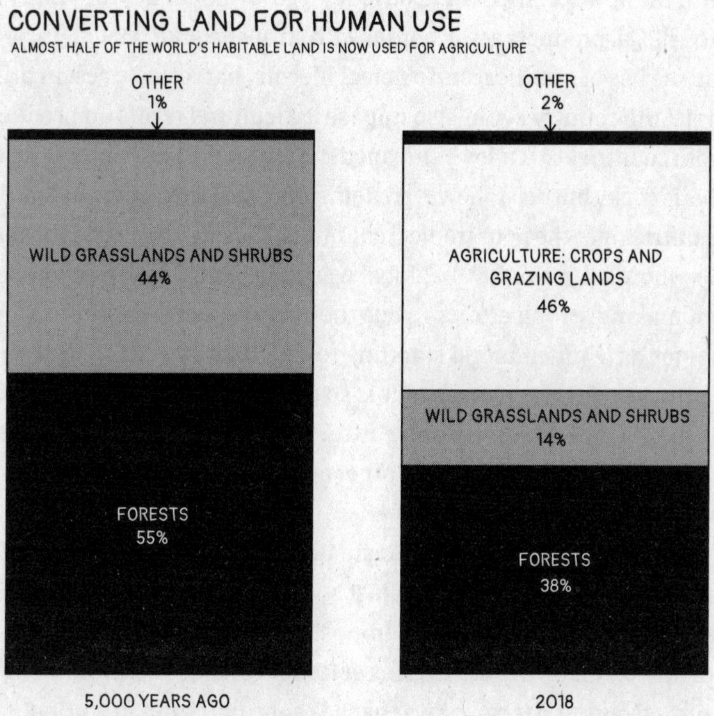

CONVERTING LAND FOR HUMAN USE

ALMOST HALF OF THE WORLD'S HABITABLE LAND IS NOW USED FOR AGRICULTURE

OTHER 1%
WILD GRASSLANDS AND SHRUBS 44%
FORESTS 55%
5,000 YEARS AGO

OTHER 2%
AGRICULTURE: CROPS AND GRAZING LANDS 46%
WILD GRASSLANDS AND SHRUBS 14%
FORESTS 38%
2018

Much of the effort to save tropical forests has come from economic pressure, not from Western governments but from environmental groups and ordinary citizens. They have targeted Western brands concerned about their image, to put pressure on their suppliers in tropical countries, demanding the adoption of sustainable production methods. It has worked, to some extent, for commodities like coffee, chocolate, and more recently, palm oil, the engine of so much destruction in Indonesia. However, the global meat industry, with far more money at stake, has largely managed to fend off such efforts. And immense smuggling networks still exist for products that are technically illegal under international law, including endangered tropical hardwoods.

We believe the United States must supplement these citizen efforts with the power of the state. The federal government has the ability to stop illegal products at the border with stronger enforcement, including the use of genetic tests to detect illegally harvested trees. In principle, the country could also impose agricultural tariffs on products from countries that allow rampant deforestation. Exactly how to do all this is tricky business, however. Remember that it is the big industrial countries, not the poor tropical countries, that are largely responsible for where we are today on global warming. Until the rich countries are making strong efforts to cut their own energy emissions, they do not have much moral standing to wag their fingers at far poorer countries. This is one reason it is so critical for the United States to develop a credible emissions-cutting program at home—until then, when Americans go lecturing other countries about their emissions, we'll come off as hypocrites.

If we are really going to tackle deforestation, it is vital to offer tropical countries carrots, not just sticks. One form those could take is trade deals: instead of just imposing tariffs on bad products, we might negotiate favorable terms of trade for an entire country's agricultural output if it can prove that it is controlling deforestation. The ultimate carrot, of course, would be cash. For well over a decade now, rich countries have been promising to pay to save tropical forests, using satellites to monitor deforestation and making cash payments to countries that bring it under control. The idea is that tropical countries need money they do not have to enforce their forest laws, to create alternative livelihoods for people who now make a living logging or subsistence farming, or to better the lives of indigenous tribes that own large tracts of forest. So far, only a few billion dollars a year are being spent on the problem, compared to the tens of billions that were promised. We think the rich countries, including the United States, must come up with ways to supply the money. It is one of the best investments the world could make, not only to save forests that are offsetting our industrial emissions but to conserve the biological richness of these places.

Unfortunately, we're pretty convinced this kind of thing can only work at the scale of entire landscapes—say, an entire Brazilian or Colombian provincial government. Funding piecemeal projects would create a real risk that deforestation avoided on one plot of land would simply be displaced to another. The governor of the province would have to be the party responsible for meeting deforestation targets; if forest loss goes down, the governor would get the promised money from abroad. These days, deforestation can be measured well using satellites, so governors who were failing would not be able to hide.

Simply stopping the loss of forests, as important as it is, will not be enough. Just as with tree-planting in the United States, tropical countries also have a major opportunity to restore forests on degraded land, or on marginal farmland. Second-growth tropical forests are already a vital part of the landscape in some places. Costa Rica is probably the world's standout example. Between the 1940s and the 1980s, the country lost 70 percent of its tropical forest cover. But then visionary leaders adopted a policy of compensating farmers who allowed the forest to regrow, financing the program with a small tax on fossil fuels. Forest cover in Costa Rica almost tripled in the decades since. Costa Rica was able to afford this program given its relatively successful tourist economy, but poorer countries will need outside help. In some places, active—and costly—tree-planting campaigns will be needed. But the tropics are so fecund that any patch of land reasonably near an intact forest will regrow on its own if farmers stop planting crops. Animals and the wind will deliver the necessary seeds. After all, forests have long recovered from natural calamities like fires and storms; they can also recover from the human calamity, if allowed to do so.

"The forests know how to do this," said Robin Chazdon, an American scientist who has spent decades studying second-growth forests in the tropics. "They've been doing it forever, growing back."

Feeding the World

Poor people all over the world were in an uproar. In Mexico City, marchers banged metal pots and pans in front of government buildings for days on end, theatrically turning the pots upside down to show they were empty. In Egypt, rioters burned cars and knocked out the windows of government buildings. In Haiti, the riots grew so fierce that the country's limited hospital beds filled with injured people, and public anger forced senators to fire the prime minister and form a new government.

Riots or public protests erupted in at least thirty countries, and tensions escalated in dozens more. The cause of all this turmoil, starting in late 2007 and cresting in 2008, was a rapid global escalation in the price of the basic food commodities, including rice, wheat, and corn, on which poor people depend. Failed harvests in several major producing countries, caused or exacerbated by the intensifying heat waves that are characteristic of the climate crisis, was a major underlying reason for the price run-up. The United States is a large grain exporter, so an American policy of turning nearly a third of the country's corn crop into ethanol for use in vehicles did not help. If we run a contest for grain between America's SUV drivers and poor people living on the margin, guess who wins.

If governments had reacted wisely, the crisis might have been manageable, but they reacted foolishly. Panicked by the prospect of running out of food, some middle-income countries started closing their borders to grain exports, even as many poorer countries were desperately trying to get their hands on supplies. The result was a predictable chain reaction in which more and more countries effectively hoarded food within their borders, escalating a global price spiral. The spikes were felt only modestly by people in rich countries, where basic commodity prices represent only a small part of the cost of processed foods. But in countries where people spend the bulk of their income on a bit of grain to get their families through the next day, the result was chaos. Over those years, global hunger made its largest jump in decades.

The global financial crisis that followed later in 2008 and 2009 helped to drive prices back down, and several years of robust harvests refilled depleted global stocks. Yet the whole affair was replayed a few years later, in 2011, after a new round of failed harvests. Grain prices spiked even higher than they had in 2008. But this time, governments were better prepared, and financial experts worked behind the scenes to head off the panicky border closures of a few years earlier. The second crisis ebbed with less public anger than the first, though it was still hard on the world's poor. In the years since, harvests have been good most years, and global commodity markets have been operating smoothly. Perilously low stockpiles have been rebuilt to a degree, and commodity prices are once again low enough that farmers in exporting countries are complaining about them.

Yet we view those events of more than a decade ago as a warning. One of the major human risks of the climate crisis—perhaps the single most important risk—is the possibility that farmers will not be able to keep up with global food demand as heat waves, droughts, floods, and other erratic weather threatens their crops. Scientists have calculated that escalating heat may already be shaving a few percentage points off the production of some major grains. That could be one reason, though probably not the only one, that the growth rate of food production has slowed compared to the halcyon years of the Green Revolution. "What a horrible world it will be if food really becomes short from year to year," said Matthew Reynolds, an agronomist. "What will that do to society?"

Few people realize how fragile the world food system really is. You might imagine that governments keep deep reserves of grain and other commodities in storage somewhere, ready to be whipped out in case of emergency. Does the United States not have years of food stocked up deep inside some secure underground storehouse? Unfortunately, the answer is no. The strategic food reserves that do exist are relatively small, and most countries have none. Larger stockpiles are held by companies in the global food supply chain, and they can normally compensate for one year's bad harvest, but not for several bad ones in a row. The human race essentially lives on what it grows from year to year, and since most of the planet's land is in the Northern Hemisphere, most of the world's food is grown there. Which means that a few bad years of weather in a row in one hemisphere can propel the whole world into serious trouble. It is not out of the question that something as simple as a large volcanic eruption, throwing ash into the air and blocking sunlight, could cause a tumultuous global food crisis.

The international trade of commodities is critical to keeping people alive. Entire countries in the Middle East and North Africa import the bulk of their food supply and have no practical alternative. The region has 400 million people now, and the population is still growing rapidly in parts of it. As we already mentioned, China is also increasingly dependent on imports, though it does produce the bulk of its food. South America has become a major grain producer, but Africa still lags badly in production per acre.

Agriculture is never a solved problem. The insects, funguses, and other pests that attack crops continue to evolve in response to farmers' efforts to suppress them. This means money must be invested continually in breeding more resistant crops just to maintain past levels of production, much less to increase them. Global agriculture has, so far, been successful at keeping up with the demand for food, even as that demand has skyrocketed. It is one of humanity's most spectacular achievements, and we say that knowing the high environmental cost of modern farming.

Past increases in demand were met not just with technical innovation, but, as we have already mentioned, by expanding the amount of land under plow. Whole forests were chopped down, and the destruction continues today. And not only forests are at risk. The international pressure that led Brazil to crack down on deforestation in the Amazon displaced the growth of agriculture into a Brazilian region called the Cerrado. A large, biologically rich landscape of savannahs and grasslands has been plowed under to create a juggernaut of an industry producing soybeans, beef, and pork, much of it exported to China.

The food system is beset by ideologues professing certainty about how things could be done better. We do not claim that kind of certitude, but we do think some things are clear. The global expansion of agriculture must stop. We cannot keep chewing up the world's last remaining wild places. The expansion onto more and more land is called agricultural *extensification*; it must be replaced with agricultural *intensification*, meaning that existing farmland yields more production per acre, year by year. That is already a well-established trend among farmers in the developed world, but less so among the tens of millions of small farmers in the developing world. The difference between what they produce on an acre of land and what they could produce is called the yield gap, and half a century after the Green Revolution, it remains enormous, particularly in Africa. Where American farmers in the Midwest typically produce 160 or 170 bushels of corn per acre in a good year, small farmers in Zambia might produce less than 50 bushels.

What do farmers need in order to pull off this intensification of global agriculture? The list is rather long, but it is not mysterious. Many farmers in Africa still do not have access to modern fertilizers, particularly the nitrogen fertilizer that is, you will recall, the main limitation on plant growth. In parts of the world, farmers lack roads, they often lack capital for buying seeds or equipment, and they sometimes lack access to basic information like prices in the markets. Developing countries need to invest more in their agricultural systems, which they did not do in the years after the Green Revolution, when global grain prices were low. They also need help: after the food crises of 2008 and 2011, the world's governments pledged billions for agricultural development, and more or less kept that pledge by investing in global research and in farm aid programs. But we fear they are losing focus again as memories of the food riots fade. International agricultural development is one of the most critical issues on the global agenda, and one of the most important steps we can take to build climate resiliency into the global economy. The American government neglected the issue in the four years of Donald J. Trump's presidency, but we hope it will become a high priority for the Biden administration. A private charity, the Bill and Melinda Gates Foundation, invests heavily in improving the lot of poor farmers, but even a charity as large as that one does not have the money to do all the work required without backing from the world's governments.

A huge contribution to solving the global food problem, and to limiting the greenhouse emissions from agriculture, would be to restrain the growth in demand. And the most effective way to do that would be to discourage the consumption of meat and dairy products. As consumers in the developing world grow richer, they are rapidly adding protein to their diets—not yet at the level at which Americans and Europeans eat meat, generally, but getting there. Several pounds of grain have to be grown and fed to animals to supply one pound of meat or dairy products, so that all meat and dairy production is intrinsically wasteful—of water, of land, and of the fossil energy that was used to grow the grain. Yet, as beneficial as it would be to persuade people—especially in the developed world—to eat less meat or cheese, it is not at all clear how to do it. After decades of discussion in the United States about the environmental cost of meat production, only about 5 percent of Americans identify themselves as vegetarians, according to results from the Gallup Poll, and only 3 percent eat stricter vegan diets. Vegetarianism may not be the answer for most people, but perhaps some moderation is in order? Almost a quarter of Americans told Gallup in 2019 that they had cut back on their meat consumption—which you can interpret as good news, but given how much meat Americans eat, it would be good for the climate and our own health if 100 percent of us cut back.

The United States government, which is heavily influenced by lobbying from the meat industry, has made little attempt at persuading people to change their diets. Nor, for that matter, have the other governments of the world. They are letting consumer sovereignty prevail, focusing on meeting the rising meat demand rather than trying to restrain it. You might say that governments are being chicken on this subject.

And speaking of chickens, they may actually be one of the answers. When Herbert Hoover was running for president of the United States in 1928, some of his supporters in the Republican Party took out a full-page advertisement in the *New York Times* with the big headline at the top "A Chicken in Every Pot." It sounds, to modern ears, like some kind of silly campaign stunt. But to families of the 1920s, that promise would have meant a great deal. In those days, cooking a whole chicken was a rare treat, something an urban family might do once or twice a year. The promise of a chicken in every pot resonated in 1928 as a promise of genuine prosperity for all Americans.

In the decades since World War II, the production of chicken has been industrialized, and the price has fallen in relative terms. As chicken has become cheaper and more plentiful, and as nutritionists have cautioned against eating too much red meat, chicken has been conquering the American diet. Beef consumption per person has actually dropped 38 percent since its peak in 1976, while in the same time chicken consumption per person has jumped 129 percent. People do not think anything nowadays about ordering a chicken sandwich for lunch or carrying a roasted chicken home from the grocery store for dinner, but those are relatively new developments.

Why, you might be asking, is it good news? The simple reason is that while all meat is environmentally costly, chicken is less so than the other common meats. The relevant comparison is known as feed-conversion efficiency: How much of the grain an animal eats does it convert into edible meat? For beef cattle, the answer is approximately 10 percent; for hogs, about 20 percent; and for chicken, about 40 percent. You might think of eating beef as equivalent to driving around in a big, gas-hungry sport utility vehicle, while eating chicken is more like driving a thrifty Honda Civic. You are damaging the planet in both cases, but if the choice is one or the other, go for the Civic. Or the chicken sandwich instead of the hamburger.

You can probably sense by now that on this subject, we believe in steady and serious incremental change. Persuading people to eat less meat, or use fewer dairy products, are steps in the right direction even if they are not ready to commit to vegan diets. We need far more experimentation with how to get people to shift their diets. One research group at the University of Oxford has proposed a meat tax to cut consumption, mainly to improve people's health but with a likely side effect of cutting greenhouse gases. No government has been bold enough to try it yet, so we really have no idea how well it would work.

Rising meat consumption has become such an urgent global problem that it is, finally, calling forth a spirit of innovation. While meat substitutes made from plants have been on the market for decades, most people found these "veggie burgers" and the like unappealing substitutes for ground beef, so the market was limited. But that is now changing fast, opening a window of opportunity for consumers to pull the American food system in a new direction.

Pulling the Levers

If you walk into any Burger King in the United States, you will find that the venerable Whopper now has competition. For about $1 more than a regular Whopper, you can get an Impossible Whopper, with a patty made not from beef, but from a mixture of ingredients derived from plants.

Ladies and gentlemen, buy that burger!

We told you at the beginning of this book that we did not think the climate could be salvaged by earnest green consumers making different buying choices. Instead, we have called for green citizenship—that is, for voters to insert ourselves into the mechanics of government to demand climate-smart choices. However, we now offer something of an exception to that rule. In the case of the food supply, we think consumer choice *can* make enough of a difference to matter, sending market signals to reward innovation that cuts emissions in the food system.

Why do we think so? To put it simply: Silicon Valley is on the case. That is, society has arrived at a moment when real innovation is starting to occur, much of it originating in the technology cluster around the San Francisco Bay area. The Impossible Whopper is a version of the Impossible Burger, which is becoming widely available across the United States, and may have already shown up in your local grocery store. It comes from a company called Impossible Foods, of Oakland, California, which has just about mastered the art of making imitation ground beef from plant ingredients like pea and soy protein, yeast extract, and several others. Impossible Foods has serious competition from a company called Beyond Meat. Both make convincing imitation ground beef. They and a wave of upstart competitors are going beyond that, though, in an effort to replace just about everything in the meat and dairy cases with plant-based substitutes. If you can't already find convincing alternatives to cheese, eggs, fish, and milk at your local grocer, you are going to be able to soon. You might already be familiar with plant-based milk alternatives, like almond and soy milk. That is now a well-established product category with close to $2 billion in annual sales, and we think additional products are going to follow the path blazed by these milk alternatives.

The big problem is that $1 premium at Burger King. Or rather, the premiums you have to pay for nearly all these products anywhere you buy them. Instead of being $1 more expensive, we think that Impossible Whopper needs to be $1 cheaper, at least. The situation at the grocery store is more dramatic still: even after recent price cuts, patties from Impossible Foods can cost twice as much at the grocery as ground beef. That needs to come way down. Lower prices for plant-based foods will drive millions of people to try them and, we suspect, to like them. By now you know the refrain: we need to buy meat alternatives down the learning curve. But, whereas consumers do not really have the ability to do that when it comes to clean electricity or steel, they do have that power when it comes to food. Just as with flat-screen televisions twenty years ago, the more people buy these products, the faster they will scale, and the faster the prices will drop.

Vegan and vegetarian diets are also good for the climate, and better for your health than eating meat alternatives, which tend to be high in fat, salt, and carbohydrates. But we live in an imperfect world, and we know that the number of Americans eating vegetarian diets is still small. Our guidance is simple: eat vegan or vegetarian if it suits you. If you can't go all the way, maybe you could become a "part-time vegetarian." Meatless Mondays are a good start. Otherwise, make climate-smart dietary choices wherever possible: pork is better for the climate than beef, and chicken is much better than pork *or* beef. Sub in an Impossible Burger or another meat alternative whenever you can. Cut dairy products in your diet to the degree possible.

In principle, it ought to be achievable to produce these alternative proteins more cheaply than meat, since the production chain for plants requires less energy, land, and water than that for animals. We think it is just a matter of scaling the market, and every reader of this book who tries these products on a regular basis will be helping to do that. Many billions of dollars of capital will ultimately be needed to bring such products to mass scale, and we think those billions will flow when investors see that the markets are growing. And they *are* growing already: sales at Impossible Foods have been rising for a decade, and some other companies in this category are showing similar gains. Let's help them grow faster.

Coming behind the plant-based method is another substitute for animal agriculture, a potential product category that would have seemed like science fiction only a few years ago. A wave of startup companies is working on growing actual meat tissue in big vats called bioreactors—essentially turning meat production into a real industry, not the half-farm, half-industrial operation that it is today. We caution that few studies are available yet on the environmental footprint of these operations, so the idea that they will be less wasteful than animal farming is still an assumption. The early research suggests that these products must be grown using renewable energy to achieve significant emissions cuts. We will soon find out what is possible. We hope readers will be open-minded enough to try these products when they come to market. That has already started to happen, with lab-grown chicken nuggets now for sale in a handful of locations in Singapore. A plate of them goes for close to $20, so it is obvious they are still poised at the top of the learning curve.

Even on the subject of the food we eat, advocates for saving the climate cannot duck out of politics entirely. The companies producing meat alternatives face a serious legal threat: a wave of state laws trying to stop them from using terms like "meat" or "burger" or "butter" on their packages. These are being passed at the behest of the meat industry, which is trying to beat back the competition while it's just getting started. More than a dozen states have already passed such laws, though they are being challenged in court as violations of the First Amendment right to free speech. We are hopeful the courts will overturn them, but we also need to fight them in every state where they have yet to be adopted. Make no mistake: these laws are not a legitimate attempt to protect consumers from confusing labels. Certainly, the words need modifiers to be legitimate, but that is enough. Nobody is going to be confused by a label that says "plant-based butter" or "veggie burger." These laws are just protectionism, and we need to raise our voices to demand that legislators reject them.

We have other levers to pull when it comes to land use. Earlier in this chapter, we mentioned "tree equity"—the idea of correcting the historical disinvestment in Black and brown neighborhoods by planting trees there. This needs to become a political target in every city in the United States. The problem is not as simple as it might seem at first glance, though. Just planting trees is not enough; they have to be maintained and, especially in the American West, they need to be watered when the ground is dry. Maintenance and watering cost money that poor families do not have, so this work will have to be done with public money. To demand that your city do that is a demand that will benefit both social justice and the climate.

Ultimately, we need Congress to act to authorize and fund tree-planting across the United States on a massive scale, and to adopt measures to encourage the conservation of tropical forests, including much tougher penalties on the importation of illegally harvested wood. It needs to double the size of the Conservation Reserve Program, as we mentioned, and expand existing programs that pay for cover crops. Until the politics are lined up for ambitious national action, states and cities should get a running start with their own pilot programs. An organization called American Forests is intently focused on many of these issues, including "tree equity" programs in poor urban neighborhoods, and we urge readers to get involved in their work.

CHAPTER 7

The Stuff We Make

The steel mill that built the West rises from the dry plains of eastern Colorado, the Rocky Mountains looming in the distance. Built in the nineteenth century, Colorado Fuel and Iron was for decades the largest mill west of the Mississippi River, controlled in its early days by the heirs of tycoons like John D. Rockefeller and Jay Gould. Its place in American history is legend: the mill churned out most of the steel rails that railroads laid across the American West, helping to transform a nation of small farms into an industrial powerhouse.

But the great mill in Pueblo, Colorado, has seen its share of misfortune, from some of the bloodiest labor strife in the country to decades of financial struggles. It has passed in and out of bankruptcy and been sold repeatedly. Today, a huge Russian steel conglomerate controls the American company that owns the mill, and the owners have cast a gimlet eye on the costs and profitability of the operation. In recent years, the Pueblo mill has said publicly that it might have to close if the managers cannot get costs down.

Enter the most unlikely of saviors: in 2019, the owners struck a deal to permit a huge solar farm next to the mill. When finished, it will be one of the largest solar arrays in the United States. Much of the power will go to the mill, which uses a huge electrical arc to melt down scrap steel for recycling into new products. Excess power will go into the electrical grid. As part of the arrangement, the mill got a break on the electric rates it pays to Xcel Energy, an essential factor in prompting the Russian owners to invest in updating the business, preserving nearly a thousand jobs. For the first time in the nation's history, but probably not the last, cheap solar power is helping to save a mighty industrial plant. And the mill in Pueblo will be able to claim that it is producing the greenest steel in North America—not only recycled but produced using solar power.

Most people go about their lives without spending much time thinking about steel—or about that other great building block of human civilization, cement. Cement is the single most ubiquitous product we make, the fundamental ingredient in nearly everything we build. Cement is a binder—a glue, in essence—that is mixed with sand, rocks, and water to form concrete. A version of the formula was first perfected in the days of the Roman Republic, transforming architecture forever in what is sometimes called the "concrete revolution." The Romans used their concrete to build roads, bridges, and aqueducts that are still in use today, and to put up a building—the Pantheon—that is topped by what is still the largest unreinforced concrete dome in the world. Steel and concrete were finally married in the nineteenth century, with an internal steel skeleton giving extra strength to large, heavy concrete structures. That combination made it possible to build bridges, tall buildings, roads, and rail lines at the immense scale that defines the modern world.

Today, the fastest trains run along steel rails mounted to concrete cross ties. We drive cars over soaring concrete bridges reinforced with steel bars. We take elevators made of steel to the tops of tall buildings made of concrete reinforced with steel. Our houses and apartment buildings sit on concrete slabs, our light poles are increasingly made of concrete, and many of our roads are made of concrete. The world produces 38 billion tons of concrete each year—more than 5 tons of concrete products for every man, woman, and child on Earth.

Unfortunately, the production of 4 billion tons of the essential ingredient in concrete, cement, is a huge source of greenhouse gases. In fact, this single industrial process is responsible for nearly 7 percent of global carbon-dioxide emissions, while producing iron and steel is responsible for close to 10 percent of CO_2 when you include the electricity both activities consume. Taken together, they are one of the largest problems we face in combating climate change. The problem is not just that iron, steel, and cement production require lots of energy from burning fossil fuels, though they do. An even bigger problem is that producing the materials involves chemical reactions that release greenhouse gases directly. The production of cement, for example, starts with cooking limestone, made primarily of calcium carbonate, into lime, or calcium oxide, in a giant kiln. The chemical reaction liberates carbon and combines it with oxygen, sending carbon dioxide into the air. This means the very sidewalks and stairs you may take to work every day contributed to the gathering climate crisis.

In this chapter, we turn to industry, the economic sector that produces the immense flows of material on which modern civilization depends. Your life might feel pretty remote from the world's steel factories, its cement plants, its oil refineries, its chemical factories, its pharmaceutical producers, and its fertilizer plants, but that is an illusion. Every product you can buy—every sack of food from the grocery, every iPhone, every car, every light bulb, every house and everything *in* the house—has come into existence only because of these material flows. Right now, essentially all of it depends on fossil fuels. If you include the electricity that factories use, global industry accounts for 30 percent of the world's carbon dioxide emissions.

We will not solve the problem of climate change without overhauling our methods of industrial production. It is one of the most daunting tasks we confront, for the scale of these businesses is hard to fathom. Some factories stretch a mile long, mines stand out against the landscape when viewed from space, and immense furnaces cook ore into iron and steel at two thousand degrees Fahrenheit. Putting up a few solar panels or buying an electric car might seem pretty trivial compared to all this digging, grinding, heating, smelting, casting, refining, and bolting together of parts. And though it has clearly begun, the hunt for alternative, low-emission ways of doing all this work is still in its infancy.

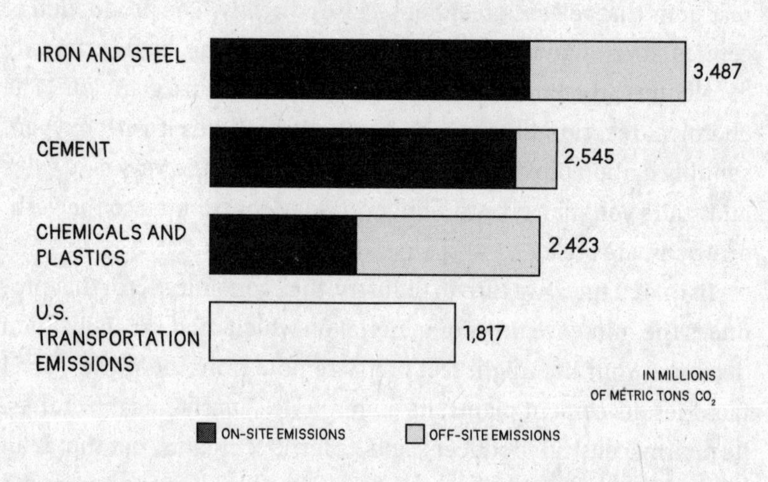

Fortunately, businesses have long had an incentive to cut emissions that has nothing to do with the climate—energy is a major cost for them, so trying to use less of it helps their bottom line. Economies around the world have been getting more and more efficient, in the sense of using less energy to produce a dollar or a yuan or a euro of economic output, for decades. Of course, overall economic and population growth have still driven emissions up, but it would have been much worse without the efficiency improvements.

That gradual ratcheting down of energy intensity is laudable, but it is not enough. We have to go much faster to solve the climate crisis, not just making existing methods more efficient, but replacing them with entirely new approaches that eliminate greenhouse emissions or at least cut them sharply. All over the world, forward-thinking business leaders are in the early stages of exploring new industrial methods. Yet many of these efforts are fragile and tentative, hobbled by the same flaw as every other attempt to solve the climate issue: the lack of strong public policy to drive emissions reductions.

In industry, perhaps more than in any other part of the economy right now, we urgently need to send signals that new, cleaner ways of doing things will find a market. Most ordinary citizens, of course, do not run cement factories or steel mills, nor do we have much say over the way our phones or shoes or sidewalks are made. What, then, can we do to pull companies in the right direction and encourage them to find ways to cut emissions faster?

How to Do It

Three approaches promise to cut industrial CO_2 emissions across the board, and we need them all. The first is to make industry more efficient, but at a faster rate than in the past. The second is to power production with different fuels. And the third is to employ improved design and better technology to slash the use of materials and of energy. Some of these approaches are likely to apply only to particular industries—using hydrogen derived from renewable electricity to make steel, for instance—while others apply broadly across many.

As we have mentioned, the search for greater efficiency is a mantra for businesses already. Sometimes the work goes quickly, with huge steps down the learning curve, and sometimes the work proceeds in only slow increments. Still, it is the engineer's creed to find better ways to do things, and free-market economies love efficiency, too, so even without strong policy, the efficiency engine labors away. The question is whether it can be sped up, so that companies can slash emissions while still earning profits.

Efficiency comes in two flavors—by device and by system. The device side is easy to understand: the best electric motors in the world, operating at optimal speed and load, convert almost 100 percent of the electrical energy supplied to them into motion. A big motor hitting this high mark can potentially save millions of dollars in electrical energy costs over its lifetime, compared to a cheaper motor with parts that rub together, creating friction and wasting energy as heat. The cheap motor might be only 80 percent efficient.

Altogether, motors consume half the world's electricity. That might seem surprising, until you realize that motors are doing the work behind the scenes in many types of equipment, from elevators to refrigerators to sliding doors. They circulate air through our office buildings and skyscrapers, and they run the air conditioners. They pump all the water and sewage in our world. They move assembly lines, run chiller plants, drive grinders and crushers, and power every type of material handling. They operate as force multipliers for pretty well every human activity.

If motors were better selected, and, more important, better operated, that consumption could be slashed. Take a simple system of a motor pumping water. In the typical setup, you turn on a pump driven by a motor, then adjust a valve to get the amount of water you want at any given time. But this approach is the energy equivalent of driving a car with your feet on both pedals. It would be pretty foolish, would it not, to have the accelerator floored at all times, while adjusting your speed solely with the brake? Yet more than 80 percent of motors are managed this way—with the electricity on full, and the work being throttled mechanically. That is an engineering embarrassment.

The better way is to use electronic controls to vary the speed of the motor itself, instead of choking back the flow with a valve. Hard as it is to believe, if applied consistently around the world, this change alone could save energy equivalent to the output of several hundred nuclear power plants. One reason it has not happened to a sufficient degree is that variable-speed motors are more expensive—but as in so many other aspects of the energy problem, the investment will pay itself back in energy savings in just a few years.

"System efficiency" takes this motor idea and extends it. The energy required for pumping depends on the amount of "pure work"—in physics terms, the amount of energy needed to overcome gravity by, for example, moving fluid several stories up in a factory—plus the parasitic drag in a particular system, such as friction in the pipes. The trade-off in this instance is rather simple: long, skinny pipes with lots of bends in them make the pumps work harder. Fat pipes with gentle bends save energy. The added energy required in a bad layout can be several times the energy required in a good layout. That's what system efficiency is all about: How can you adjust any and all elements in a system to make them more efficient?

The problem is that system efficiency is a complex, and thus underused, engineering practice. Most engineers use standards codified in handbooks for specifying motor size, valve type, pipe size, and so forth. They lay out the pipes after first placing all the machinery, with a resulting jumble of sharp bends and long stretches. They use mechanical throttles because that is easy, standard practice. All of this can multiply the energy use in a factory, and thus the greenhouse emissions, by as much as fourfold over what is strictly necessary.

Governments are understandably reluctant to interfere in complex engineering decisions like factory design, as well they ought to be. But they need to push their industries to go faster in reducing emissions. What can they do? One basic strategy is that every major category of industrial device should be subject to a legally enforceable "minimum performance standard." Every motor, air conditioner, compressor, pump, and so forth should be built to perform at top levels of efficiency. And those standards should be ratcheted up every few years, so that continual improvement becomes a habit in the engineering professions. This standard-setting approach has been used in most industrialized countries, including the United States, for decades, and has delivered substantial results. But the process of setting and upgrading those standards has been agonizingly slow, even in the countries that do it regularly, and each increment is often more modest than it should be. A big reason for this is one we've already encountered: the buyers of industrial equipment are just as sensitive as the buyers of cars to the sticker price, never mind the long-run savings. So the companies that make the old equipment fight higher standards. Collectively, this means we have left billions of dollars on the table, and emitted billions of tons of carbon dioxide into the atmosphere, unnecessarily.

A broader type of performance standard can also be used for entire industrial sectors—such as the cement industry or steel-making. In this approach, the policy is to set a maximum level of emissions per ton of concrete, say, and edge that number down every few years. Plant owners will likely complain that the public officials setting the standards go too far, impose excessive costs, and are arbitrary. These complaints are potentially legitimate, but there is an almost magical way around them. The government can look at the best-performing *existing* cement or steel plants for proof of what is possible, and make that the new minimum for all plants—then repeat the exercise every four or five years. The top standard in efficiency today becomes the new minimum tomorrow.

This idea was pioneered by the Japanese government, and it is called the "Top Runner" system in that country. The Japanese use this for a huge variety of products, from consumer goods to heavy industry, and the system has helped to make that country a world leader in energy efficiency.

One great feature of setting an emissions standard for every ton of output is that it leaves the particulars to plant engineers. This avoids the problem of having government bureaucrats crawling around factories making decisions that ought to be left to professionals. The engineers are simply given a target, and they have all sorts of options to meet it: they can refit furnaces, capture and reuse waste heat, change pumps and piping, think of new chemistries, look for cleaner sources of energy supply, and so forth. The point is that a continually escalating standard taps their skills and expertise and turns them into advocates for a cleaner industry, rather than nodes of resistance.

Another option is for governments to impose a serious price on industrial emissions of carbon dioxide and other greenhouse gases. This can be a tax, or it can be a requirement to buy pollution permits that are sold in a marketplace. Elsewhere in this book we have expressed skepticism about the idea of putting a price on emissions of carbon dioxide and other greenhouse gases, largely because it has been politically difficult to pass such policies at a price high enough to make any real difference. But if those political problems could be overcome, industry is the sector where we think a carbon price would do the most good. Set high enough, it would alter the thousands of economic calculations that any large business has to make every week, especially the companies with the highest emissions: steel, cement, chemicals, and so forth.

Companies would also need to know that the carbon price is going to rise over time, so that early investments in cutting emissions will pay off. California has come up with one way of doing this, a "cap and trade" program that requires big industries to purchase permits from the state granting them the right to emit greenhouse gases. Small businesses are exempted. The state essentially puts a floor and a ceiling on permit prices, offering businesses some certainty, but it expects to sell fewer and fewer permits as its emissions cap tightens. That means the prices should rise over time. It gives industry the time and the certainty to work toward clear goals. And because trading of permits is allowed, the system can help find the cheapest ways to cut emissions across the California economy. That is because the companies with the cheapest methods to reduce emissions can "overcomply," selling their excess permits in the marketplace to companies that have trouble cutting emission. This lowers the cost for all companies. We emphasize that these types of programs are no panacea, however. They require sophisticated monitoring and auditing, or companies will be tempted to cheat.

The difficult politics of taxing emissions can lead to start-and-stop policies that create huge uncertainty for industry. Australia, for instance, imposed an emissions price in 2012 and then repealed it two years later when the government changed hands, shattering business confidence in the policy's long-term direction. Industrial equipment lasts decades, sometimes half a century. The best time to make it efficient is when it is first being built, or substantially upgraded. Improvements then are cheap or free; retrofitting a plant early in its useful life is expensive. Only in the Nordic countries have governments been sending consistent signals for a long time about cutting emissions. In the rest of the world, political failures and uncertainty have led to carbon emissions from industry remaining higher than necessary, with too little investment in clean manufacturing.

New Fuels, New Chemistry

Another way to clean up industry is with new, lower-carbon fuels. To some extent this shift is happening automatically: remember that a lot of the energy used in factories is electricity, and as the electrical grid gets cleaner, factory emissions will go down without the factory managers needing to do any additional work. Running the electricity system on low-carbon sources pays vast dividends in every sector, emphatically including industry. Unfortunately, a great deal of industrial emissions come not from electricity use but from the direct combustion of coal, oil, and gas as products are made.

Let's return to the example of cement. The standard production method requires burning fossil fuels, often coal, to cook limestone into lime, the main ingredient in cement. The fossil fuels produce less than half the emissions, though. The rest come from the chemical reaction that transforms limestone into lime, releasing carbon dioxide. The broad term for emissions of this type is "process emissions," and in some industries they are much larger than the emissions from electricity use.

In the case of cement, it is possible to make the heating stage more efficient with improved kilns. The formula for making concrete can also be tweaked to economize on the use of cement. But the greatest opportunity is to significantly change the chemistry of cement production. Some estimates suggest that an intensive focus on cleaning up the cement industry could cut its emissions in half—and remember that this is one of the largest sources of industrial emissions worldwide. It is a profound opportunity, and one on which we think both government and industry should be pushing much harder.

Several companies are working on lower-emissions cement. In New Jersey, a company called Solidia is commercializing a chemical formula developed at Rutgers University that has been found to cut emissions from cement production by more than half. The process has mostly been applied so far to concrete items that can be made in a factory—pavers for driveways and patios, light poles, railroad ties, and so on. Items like that make up about a third of the market for cement. The larger part of the market is for cement that is poured on-site, to construct buildings or roads or bridges. Solidia has begun to commercialize a reduced-carbon version of pourable cement. Another company, CarbonCure, is pursuing a somewhat similar approach. The good news is that the world's major cement manufacturers are showing interest in these methods, cutting deals in the last few years to try them out.

Likewise, in the production of iron and steel, a promising new approach is under development that could largely eliminate emissions. That involves burning hydrogen as the energy source to purify iron ore, instead of burning fossil fuels. The hydrogen would need to be made by using renewable electricity to split water into its constituent elements, hydrogen and oxygen. A pilot steel mill capable of working this way has gone into operation in Sweden, and Germany is pursuing the approach, as well. We will have more to say in the next chapter about the potential role of hydrogen in the future economy.

In both the cement and steel industries, alternative methods have been slow to take off—a reflection not just of the conservatism of those industries, but of the lack of signals in the marketplace that lower-emissions products will be valued. Nor have any government mandates compensated for that market failure. And that, of course, traces back to the overall failure of governments to set strong policies on climate change.

The Power of Design

We mentioned earlier that the energy needed to produce a given level of economic output has been falling for decades, offset by growing economies and growing populations. You might imagine this means waste has been systematically squeezed out of the economy. True that is, but we are sad to say an awful lot of waste remains.

The way society uses all that material streaming out of steel mills and cement kilns is a prime example. It is easy to build big, blocky, rectangular things out of concrete, because boxy molds into which the concrete can be poured are handily made by linking up wooden forms. This does not mean all that concrete is really needed—quite the contrary. The stresses and strains within those big rectangular structures are by no means uniform, which implies that for many uses, most of the concrete is just acting as filler. If it were easy to put concrete only where it was needed, drop-by-drop, then huge amounts could be saved.

A new technology has come onto the scene that could alter the way we use concrete and other materials: three-dimensional printing. You have probably seen 3D printing on a small scale in a store, with a robotic printer making trinkets by depositing melted plastic one tiny drop at a time. At that scale, the technology has already found wide application. Architectural firms use 3D printers to make models of their designs. Companies use them to make plastic parts that are needed only in small quantity. During the coronavirus crisis, intrepid hobbyists in Italy and elsewhere used 3D printers to create plastic medical parts that were in short supply.

But we are only in the early stages of discovering what this technology can potentially do. Much larger 3D printers are already being tested, capable of building complex designs by placing material drop by drop. Entire houses and offices might one day be built this way, but closer to hand is the possibility of building bike bridges, car bridges, and public restrooms. There are prototype printed buildings already. Instead of putting up huge, temporary forms into which to pour wet concrete, the new machines whirl about, putting concrete down in discrete drops or rivulets—and only where it is needed to ensure the structural integrity of the object. This kind of printing, if you want to call it that, could turn concrete from a bulky, generic material into a precision material. In essence, 3D construction, combined with some creative thinking, will enable engineers to substitute strong design for strong material.

The true advantage is that 3D printing makes complexity easy. Shapes that would be impossible, or too costly, to create become simply a matter of programming. There is a considerable distance between these early demonstration projects and a new way of doing business across the globe, of course. And 3D printing itself could potentially consume a great deal of energy—if you are melting steel ingots to deposit drop by drop, imagine the heat that will be required, so it would be important to power that machine with clean electricity.

On balance, we think the technology offers likely benefits: if it becomes possible to replace big, bulky objects with elegant, graceful objects, the potential for emissions savings is immense. Consider what it will mean for cement emissions, for instance, if every new bridge in the world requires only half as much concrete as today's designs. We think both government and industry should be pushing this technology harder in order to find out what it can really do.

The old economy—the economy of waste, of excess, of over-design, of profligacy—must give way in the twenty-first century to a new economy in which every decision to use materials or energy is carefully considered. The work must be carried out largely by private industry, but it will not proceed at the speed required until we adopt public standards that call forth the great inventive minds of the age.

Pulling the Levers

We have already mentioned what we see as the biggest problem in the marketplace: industry is not receiving much of a signal that new, cleaner ways of making products will be valued. By now, you know that all new technologies cost more in the beginning. The lack of interest in buying clean products is holding back investment in new ways of making them, which prevents these methods from jumping onto the learning curve and falling in cost.

We propose a solution to this problem: new policies requiring that governments "Buy Clean." What we mean by that is that when governments themselves go to market, offering companies contracts to install sidewalks, build bridges, supply new cars, and so forth, they ought to state a preference for products and materials with low emissions. Bidders willing to supply these cleaner products would be paid a premium. The extra cost cannot be unlimited, of course—no government wants to sock the taxpayers by turning a $1 billion bridge into a $2 billion bridge. But governments ought to be willing to pay a few percentage points extra to get materials that are safer for the climate.

Buy Clean is a potentially powerful lever, for a simple reason: governments are huge buyers of bulk materials like steel and concrete. After all, it is usually governments that are contracting to build roads, bridges, rail lines, sewer pipes, and other heavy infrastructure. In the United States, 40 percent of concrete and steel go to such government-financed projects. Governments at every level buy these materials. Both cities and states oversee road, highway, and bridge projects. The federal government finances many of the more significant projects by sending large sums of money to the states. It also spends huge sums on federal buildings.

Our political target is clear, then: we need Buy Clean policies adopted at every level of government. This idea is in its early stages, but has already started to happen. California passed its first Buy Clean law in 2017, and Colorado, Washington State, and Minnesota have followed suit recently, often starting with pilot projects. Proposals are pending in other states. And local governments are looking at the issue, too, with a handful already adopting their own Buy Clean policies. At a major climate conference in Glasgow in late 2021, the United States helped put together a global coalition of countries committed to pursuing Buy Clean policies. Citizens in all those countries need to push their governments to follow through, sending an international market signal.

Unfortunately, in California, the cement industry got itself exempted from the initial law through last-minute maneuvering. But the state legislature has since passed a measure that should, in time, bring cement under the Buy Clean requirements. In other states where Buy Clean measures pass, cement needs to be included from the outset.

What can you do as a citizen? Raise the idea of Buy Clean legislation with every politician who asks for your vote. Bring it up at your local city council or county commission meetings. While not many Buy Clean laws have passed at the local level yet, we think that is a coming wave—so we urge you to use your voice to help hurry it along.

If you work for a large corporation, there is another way you can help push to clean up industry. More and more employees are starting to use their voices within their companies to urge those firms to commit to serious climate action. Companies of every kind need climate plans. Even those that do not produce industrial goods directly are almost always *buyers* of such products. Pressure from them to clean up the manufacturing chain can make a difference.

In May of 2019, 7,600 Amazon employees signed a demand that the company develop a more serious climate-action plan. After months of internal tensions on the issue, chief executive Jeff Bezos made several dramatic commitments late that year. The company ordered one hundred thousand electric trucks to arrive over several years, part of an effort to cut emissions in its vast delivery operations. The company also enlisted other big corporations in an initiative it dubbed the Climate Pledge, asking them to join Amazon in promising to meet the goals of the Paris Agreement a decade early, by 2040. More than two hundred large companies have since signed, including Best Buy, electronics giant Philips, IBM, and Procter & Gamble. Mr. Bezos followed up these initiatives in 2020 by pledging $1 billion of his fortune to create the Bezos Earth Fund, a philanthropy to invest in climate solutions.

Employee pressure has been a vital factor inside many other companies, especially in the technology industry. Virtually all companies in that industry are now buying renewable energy to power their operations, and many are working to clean up their industrial supply chains. The movement is spreading beyond tech. An organization called Climate Voice gives specific tips and advice about how to take action inside your company. It also has an extensive outreach to college students, asking them to sign a pledge to ask questions about a company's climate efforts as they are interviewing for jobs, and to keep speaking up once they join the corporate world. We urge every college student in America to sign that pledge. Another climate organization, Project Drawdown, has published a guide for taking action from inside corporations.

We believe the steps we have outlined can jump-start progress in the industrial sector, pulling cleaner production technology into the marketplace and launching it down the learning curve. Governments need to start with Buy Clean policies, but as soon as we all get a clearer fix on how fast the supply chains can be cleaned up, they need to be prepared to leap beyond that policy. Just as they have done with consumer goods like refrigerators and televisions, they will need to impose performance standards and emissions targets on cement, steel, chemicals, fertilizer, and other elements of heavy industry.

We must arrive, and soon, at a world where cleaning up is not something industrial corporations pledge to do out of good citizenship; it needs to become the law.

CHAPTER 8

Inventing Tomorrow

Not long ago, three strange blue cubes appeared in a neighborhood of Rotterdam, the big port city in the Netherlands. They were roughly the size of shipping containers, with pipes on top, but nothing else about them offered a clue as to their purpose. Passersby might have imagined that aliens had dropped some exotic new piece of technology into the picturesque Rozenburg neighborhood. In fact, the blue boxes use renewable electricity to break water apart into its components of hydrogen and oxygen, then send that "green hydrogen" to an apartment building nearby where it is burned to fire boilers to heat part of the building. Until a few years ago, the building had used natural gas for heat.

Five thousand miles away, a much bigger industrial project has popped up on the landscape of Texas. It is an unusual new type of power plant, one that burns gas, but then captures the carbon emissions so they can be buried underground. Plans are underway to build an even bigger plant using this technology in the United Kingdom, with the emissions stored beneath the North Sea.

And in the far Northwest of the United States, planning has entered the late stages for a new type of nuclear reactor. It is meant to be safer than conventional designs, and also more modular, so that many of the components can be built in factories—potentially overcoming the costly construction snafus that bedeviled older nuclear projects. Hopes are running high that this project, being developed by a company in Oregon for a site in Idaho, can help relaunch the American nuclear-power industry.

None of these three technologies is ready for widespread use today. It is hard to say whether any of them will be practicable or economical in the future. But they are all noteworthy as efforts to invent that future—to create the technologies society will need to deploy on a large scale in the 2030s and 2040s to drive emissions down.

In this book, we have made the case that technologies already available in the 2020s can take us a long way toward meeting our climate goals. We need to push much harder on renewable energy, on electric cars and buses, on full electrification of buildings, and on repairing our cities. But this push, critical as it is, will not be enough to secure a reasonable climate future. Remember that we can only tackle the problem by cutting emissions all the way to zero—and for some major economic activities that rely on fossil fuels, we simply do not know how to do that yet.

What do we do about airplanes, for example, whose emissions are fairly small now but rising fast? Short routes may be candidates for battery-powered planes, which are under development, but long ones will continue requiring liquid fuels. How about heavy shipping, which uses large amounts of fossil fuels? In a previous chapter, we discussed the emissions from production of bulk materials like steel and cement, and the tentative efforts to reduce them—but those efforts are still far from solving the whole emissions problem in these industries.

While the crisis that confronts society requires pushing existing technologies like solar and wind power as far as they will go, the situation also demands innovation. We must create new technologies that can supply society with the goods and services that people require while slashing emissions. All over the world, engineers and companies and universities are already on the case. But at a laboratory scale, the research is going too slowly. Even worse, the early demonstration projects that can prove a technology works are moving at a snail's pace. The fundamental problem is money, of course. For energy technologies, given their immense scale, a single pilot project can cost a billion dollars or more. Private companies will often put up a lot of the money, but to take a big risk on unproven technology aimed at meeting social goals, they usually demand some help from the government.

In the United States, until just recently, the government was spending around $10 billion per year on energy research, and less than half that was being spent on projects that we would classify as likely to produce genuine innovation. This was a national embarrassment: the American people were spending more money every year celebrating Halloween than they were on energy innovation to save the planet. The situation changed dramatically early in Joe Biden's presidency, as Democrats pushed two major spending bills through Congress. Those bills contained billions for climate and energy research, as well as large tax credits for the deployment of new energy technologies. That new money should lead to some major tests of promising technologies. Unfortunately, the bills went through with minimal Republican support. Washington's newfound interest in energy innovation is vulnerable to cutbacks in a future Republican administration.

Heavy investment in innovation by the rich countries of the world is a moral imperative. Europe and the United States, the richest regions, have been founts of innovation since the Industrial Revolution. These regions got rich in large part by burning fossil fuels and they are largely responsible for the situation we find ourselves in today. That means the United States and the European Union, along with other rich countries like Japan, Canada, and Australia, must not only cut their own emissions, but they need to take the lead in developing new technologies that can be applied worldwide. It is the rich countries of the world that have both the technical prowess and the money required.

The Biden climate bills are just a down payment on what we really need. American citizens must demand that their government triple or quadruple its investment in innovative energy research, and that it urge other rich countries to do the same. The plan needs to include more funding for university laboratories where the earliest stages of research occur. But, just as importantly, the government needs to help move those laboratory innovations to commercial scale. That means placing big financial bets on first-of-a-kind power plants and other large, costly projects. Taxpayer money is precious, and of course it must be used wisely. At the same time, if none of these big projects fails, it will mean the government and the innovators it has chosen to back are not being bold enough. Some of the projects inevitably *will* fail—that is the very reason private industry will not dare do the job by itself. Sometimes, enough money may be on the line in these first-of-a-kind projects that failure could bankrupt a company.

In undertaking this innovation push, governments should be guided in particular by the idea of technological learning rates. The ideal candidates for a rapid development program are clean technologies that have just started to get cheaper, and the goal would be to accelerate that price decline until they become affordable. Similarly, brand-new technologies should receive grants and other financial support when they are still costly, but only long enough to find out if they hold the potential for big cost declines. The losers need to be abandoned early, and taxpayers' money spent on further development of the likely winners. As a matter of course, government grant managers need to be asking the question we have asked repeatedly in this book about new technologies: Have they jumped onto the learning curve?

The technologies likely to need development work number in the thousands, as nobody in the 2020s can really predict what society will need in the 2050s or the 2080s. But some technologies that warrant a major push in the next five to ten years are already on the horizon. The most essential of those are technologies that can potentially complement renewable electricity, balancing out the fluctuations from wind, solar, and hydro power. We will need these grid-supporting technologies no later than the 2040s, and perhaps earlier, to meet our climate goals. The rest of this chapter outlines four of the most promising possibilities in this area. Each of these will require intensive collaboration between government and industry.

Hydrogen: From Hype to Promise

For decades now, the idea of a coming "hydrogen economy" has shimmered in the distance, like a desert mirage. Entire books have been written about the miracles that supposedly await us when we start consuming hydrogen instead of fossil fuels. At least two distinct waves of enthusiasm for, and investment in, hydrogen have occurred since the oil crises of the 1970s. Governments have already invested billions in the prospect of a hydrogen economy, especially the government of Japan, which saw a chance to be at the forefront of a futuristic industry. Cars, trucks, and buses that use hydrogen are available today, most of them developed in Japan. Yet both of these investment waves produced too little hydrogen and too much hot air.

The obstacles to using hydrogen as a fuel were, and still are, considerable. Those hydrogen cars are not selling well, and if you do lease one in the few jurisdictions where they are available, like California, plan on encountering some serious logistical difficulties, including sometimes waiting in long lines to fuel up at the handful of stations that dispense hydrogen. Trying to take a long trip in one of these cars would entail a real risk of running out of fuel.

The earlier hype always had a grain of truth. Hydrogen does have some near-miraculous traits, and it almost certainly has a critical role to play in cutting some of the fossil-fuel emissions that are otherwise hard to eliminate. This fact has dawned recently on many governments and industry leaders, including ones that took a skeptical stance toward hydrogen in the past. We are seeing stepped-up investment, and commitments to move forward with demonstration projects using hydrogen for various purposes. The International Energy Agency recently reported that seventeen of the world's governments have developed hydrogen strategies, and another twenty are working on strategies.

Why is hydrogen such a potential miracle fuel? Basically because of its simplicity. It is the lightest element in the universe, with most hydrogen consisting of a single electron spinning around a single proton. If you burn it, two hydrogen molecules combine with a molecule of oxygen in the air to yield a burst of energy and a new compound formally known as dihydrogen monoxide—that is, water. Starting with such a simple molecule as hydrogen, you don't get as many of the nasty pollutants that come from burning more complex fuels. If you had a city with nothing but hydrogen-burning cars and power plants, air pollution would drop radically. If we could do it the world over, millions of lives would be saved every year just from cleaner air. You don't even need to burn hydrogen to use it, though. Devices called fuel cells can combine hydrogen with oxygen from the air to produce electricity and heat, without actually burning the hydrogen. The hydrogen cars and trucks that are available are really electric vehicles powered with hydrogen fuel cells. The only emission from their tailpipes is water vapor.

The trouble with hydrogen is that, in an important sense, it is not really a fuel at all. Experts refer to it, instead, as an "energy carrier." Hydrogen is so reactive with other elements that it does not exist in a free form on the Earth, at least not in appreciable quantity. That means you cannot go dig hydrogen up and sell it to people, the way you would coal or oil. If you want to use hydrogen, you have to make it by breaking apart some chemical compound, usually water, which requires an input of energy from a different source. If fossil fuels supply that energy, you may have gained nothing as far as overall emissions. So, if we are going to make wide use of hydrogen, we need another way to produce it.

Hydrogen has some disadvantages, not least that it is explosive, but also one huge advantage: you can store it. Modern, high-strength tanks and pipes can generally store and transport it safely. And that ability to store hydrogen is one of the keys to its potential role in the energy transition. Recall the argument often used against renewable energy: that the sun does not always shine and the wind does not always blow. We have already shown that this argument is exaggerated and is no real barrier to much wider use of those resources in the near term. Yet it is true that in the long term, we do have a problem. Many studies argue that the electricity system could use wind turbines and solar panels for as much as 70 or 80 percent of the supply while keeping costs reasonable, but getting beyond that is going to be difficult. The reason is that the electricity system may have to operate for days on end with little or no renewable-energy production. In the short run we are using power plants burning natural gas to fill the gap, but in the long run, we need something else.

One possibility is to store electricity when we have too much of it, then call on that supply when we are running short. Batteries can potentially do that for a few hours, but we may need days or even weeks of supply, and even decades from now, batteries could well be too expensive for such long-term storage. Instead, we could use excess electricity to break apart water molecules, pumping the hydrogen into storage in salt caverns or other large reservoirs. People have been creating hydrogen with this procedure, known as electrolysis, for well over a century. You may have even done it yourself in high-school chemistry. If we create hydrogen this way on a large scale by feeding renewable energy into devices called electrolyzers, which use the energy to split water molecules into hydrogen and oxygen, we could store the hydrogen and then use it in power plants to make up for electricity shortfalls when they occur.

Millions of tons of hydrogen are made around the world already, every year. The compound is widely used in various industrial applications, particularly chemical production and oil refining. But industry, by and large, does not make hydrogen by electrolysis. Instead, companies start with natural gas, which consists mostly of methane—a chemical compound in which one carbon atom is bound to four hydrogen atoms. The production method essentially strips off the hydrogens and ejects the carbon, in the form of carbon dioxide that is pumped into the atmosphere—yet another contributor to the problem of global warming. Partly because industry gets to dispose of its waste CO_2 for nothing, the process of producing this "gray hydrogen" is cheaper than electrolysis would be. If industry were forced to convert to electrolysis overnight, the cost of producing hydrogen might easily quadruple.

Yet recently, some of the equipment required for electrolysis has been falling in cost. These are still early days, but a frenzy is underway as universities and startup companies try to make the procedure still cheaper and more efficient. Electrolyzers have jumped onto the learning curve—their costs are falling about 16 percent every time cumulative production of the devices doubles. The other crucial piece of the puzzle will be an abundance of cheap electricity to run the electrolyzers, and with the falling costs of wind and solar power, the prospects are good. If cheap hydrogen made in this way—"green hydrogen," as it's called—becomes widely available, electricity storage is only the beginning of what we might be able to do with it.

For example, green hydrogen might be the only realistic way to eliminate emissions from certain kinds of industry. In Chapter 7, we mentioned the likelihood that this will be the best way to clean up the production of iron and steel. The chemical industry is a big emissions source because of its reliance on natural gas. The production of nitrogen fertilizer, critical to feeding the world, also requires huge amounts of natural gas. Because hydrogen can play the same role in these industries that fossil fuels are playing today, they could all potentially be made climate-friendly with an abundant supply of hydrogen.

What's more, hydrogen gas can be chemically transformed into liquid fuels that are much easier to transport and handle than gaseous hydrogen itself. One example is ammonia, which is used as fertilizer but can also be burned as a fuel. In principle, hydrogen could also be converted into other liquid fuels that closely mimic jet fuel or diesel and could go straight into the existing fleet of airplanes, trucks, and ships. The core technology needed to do this has existed since the 1920s, and was used to supply Germany with liquid fuels during World War II. But again it is a matter of cost. Today, fuels made this way would be several times as costly as ones made from fossil fuels. The situation demands focused investment from governments to drive costs down. And the critical first step is to set more ambitious targets for renewable energy, to create a supply that can be used for the production of hydrogen. Without a doubt, some of today's hydrogen hype will die down as the real economics of producing and using it at scale are discovered, but for some purposes clean hydrogen may well prove to be essential.

The Nuclear Conundrum

The cascade of failures started on the day shift, when nuclear technicians trying to clean out a filter accidentally blocked an air hose. They went home that evening with no clue what they had set in motion. Before sunrise, that seemingly small error would mutate into one of the most important events in the history of nuclear power.

It was four in the morning on March 28, 1979, when things really started to go downhill at the Three Mile Island nuclear plant in central Pennsylvania. The blocked air hose caused pumps to shut down, so that the nuclear fuel at the heart of the reactor was no longer being cooled properly. Safety circuits sensed the unusual conditions, and within eight seconds, the reactor "scrammed": control rods automatically plunged into its core, and the nuclear chain reaction that was splitting uranium atoms to produce heat came to an immediate halt. But by then, the earlier errors had put a cooling loop out of commission, and heat continued to build up after the shutdown, with gases ultimately escaping the containment dome and carrying small amounts of radioactive material into the atmosphere. Within days, 140,000 people had fled central Pennsylvania in panic. By chance, the movie *The China Syndrome*, which dramatized the potential consequences of a runaway nuclear meltdown, had opened in theaters just two weeks before. Seeing a terrifying movie appear to come to life in front of their eyes, the American people grew far more suspicious about the safety of nuclear power. The government's management of the nuclear industry came under fire.

The consequences of that series of events would play out for decades—indeed, they are with us still. Later nuclear accidents, at Chernobyl in what was then the Soviet Union and at Fukushima in Japan, cemented the opinion of many people that nuclear power was not safe. In all three cases, the calamities could easily have been much worse. The Fukushima accident, for example, came close to forcing an evacuation of Tokyo, the largest metropolitan area in the world. Still, the truth is that far more people have died from the burning of fossil fuels than from nuclear accidents—the worldwide toll from air pollution alone is calculated at more than 8 million premature deaths a year, while the highest credible estimates of the nuclear death toll put the figure at fewer than 5,000 in the entire history of the technology. No one died at Three Mile Island.

In principle, nuclear power—despite its real problems—is one way to replace the dirty, deadly black fuels that we dig out of the Earth. Yet in reality, nuclear power is a vexed and vexing technology. Figuring out what role it ought to play in the energy transition is one of the toughest problems governments will confront in the coming years.

For decades after Three Mile Island, no new nuclear plants were ordered in the United States. In the country that had done more than any other to develop nuclear power, the technology began to atrophy. The same thing eventually happened across much of the Western world. The supply chains needed to build nuclear plants deteriorated, and critical knowledge and experience were lost.

Over the past fifteen years, governments and corporations have made arduous efforts to revive the nuclear industry. In the United States, one new plant—built on a 1970s design—has come into operation under the auspices of the Tennessee Valley Authority. But a plant in South Carolina was canceled after wild cost overruns, leaving electricity customers in that state saddled with an $8 billion hole in the ground. Two reactors are under construction in Georgia, but they have also accrued billions in cost overruns. The story is the same in Europe, at Flamanville in France and Olkiluoto in Finland: years of delay, billions in cost overruns, and uncertainty about whether Western nations can ever relearn how to build these power plants on time and on budget.

Even if a handful do get finished, current estimates suggest the energy they produce will be four to five times as expensive as an equivalent amount of renewable power, and that multiple keeps rising. Nuclear power plants do offer two major advantages over renewable energy, in that they can operate day or night regardless of the weather and they can produce a large amount of power on a relatively small amount of land. But with costs ballooning on every project, investors have become extremely wary of financing these plants. The few that are being built in the West have required large loan guarantees or government subsidies, and—unlike the subsidies for renewable energy—there is no evidence those subsidies are leading to cost declines over time. Nuclear power, in other words, has never jumped onto the learning curve.

The core problem is that the current method of building nuclear plants is fundamentally prone to failure. They are huge, complicated infrastructure projects in which heavy components made of concrete and steel must be built to exacting specifications. Safety requirements can change rapidly, as they did across much of the world after

Fukushima. All of this complex work has to be done on-site, by crews that have likely never built a nuclear plant before.

This inability to build nuclear plants on time and budget seems to be primarily a problem of the Western world. The nuclear situation looks different in Asia, where South Korean, Chinese, and Russian power companies have managed to build nuclear plants at a reasonable clip and without huge snafus. The key is that contractors there are building many plants, so their teams can move from one project to the next having learned from past experience. That makes these plants cheaper than the troubled plants in the West, although still more costly than renewable energy. And even in Asia, these plants are not being built anywhere near fast enough to keep up with the rising demand for electricity. That means nuclear is losing ground as a share of the global power mix, and has been since the 1990s. What's more, the safety of the recent Asian plant designs has not really been tested—a Chinese replay of the debacle at Fukushima might radically change public opinion in that country. But for now, the plants continue to be built.

The United States could probably do what China and Korea are doing—build enough plants to regain the capacity to do it competently. But we suspect—and this is just a guess—that taxpayers would have to be willing to write $50 billion or more in checks to cover cost overruns before the industry got back to a minimal level of competence. With large wind farms and solar plants routinely going up in a matter of months, on time and under budget, you have to ask if restarting a nuclear industry plagued by outdated plant designs is really the best way to spend $50 billion on the energy transition.

To be clear, nuclear power is an important source of electricity. More than 450 plants are operating worldwide, a quarter of them in the United States. They supply about 10 percent of global electricity and approximately 20 percent of the power in the United States, making them this nation's largest source of low-emissions electricity. Even if we do not know how to build new plants, we cannot afford to lose the old ones anytime soon.

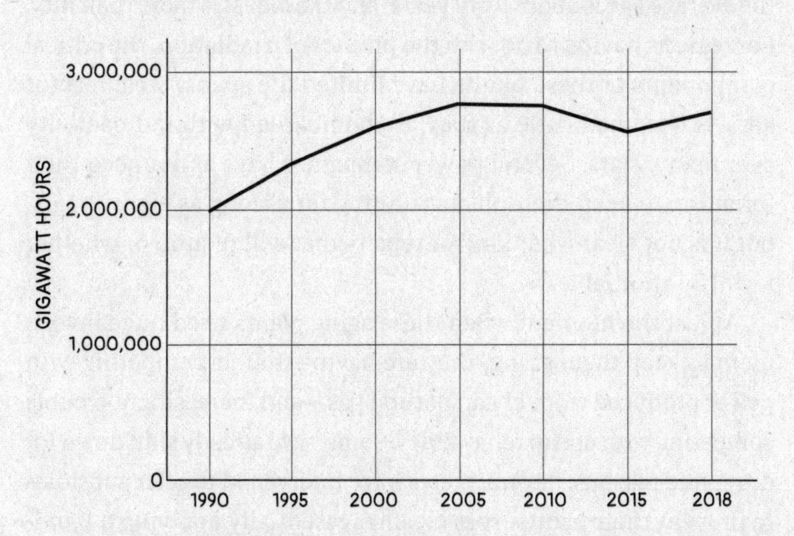
IN ABSOLUTE TERMS, GLOBAL ELECTRICITY FROM NUCLEAR REACTORS HAS RISEN IN RECENT YEARS...
GIGAWATT HOURS
3,000,000
2,000,000
1,000,000
0
1990
1995
2000
2005
2010
2015
2018

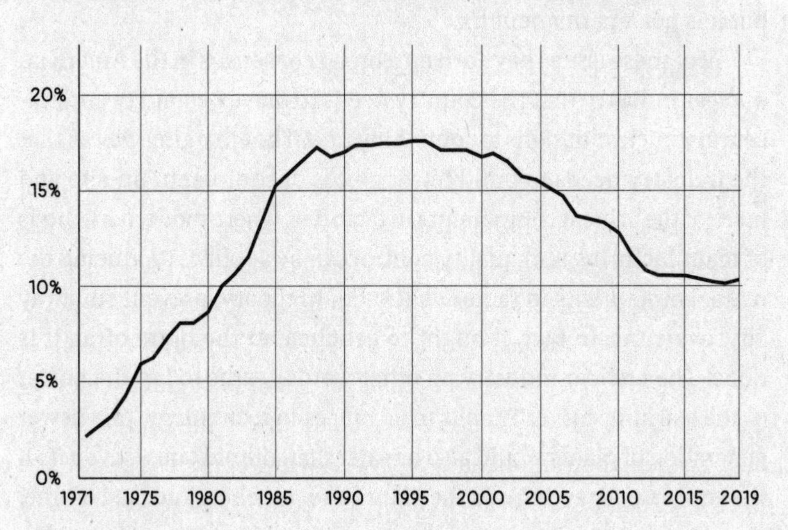
...BUT IN RELATIVE TERMS, NUCLEAR POWER'S SHARE OF ELECTRICITY PRODUCTION IS FALLING
20%
15%
10%
5%
0%
1971
1975
1980
1985
1990
1995
2000
2005
2010
2015
2019

Many of America's plants were ordered before Three Mile Island and completed in the years just after, so they are long in the tooth: the average age is about forty years, and the oldest is more than fifty. For reasons having to do with the physics of irradiation, the critical components of these plants have limited life spans. Steel reactor vessels become brittle as they are bombarded with radioactivity over many years. Several power companies have announced their intention to keep their plants running for as long as eighty years, but it is not clear what kind of repairs that will require or whether it will be affordable.

At just the moment when these aging plants need huge investment to keep them going, they are having trouble competing with power produced from cheap natural gas—and increasingly, trouble competing with cheap renewables. Some have already shut down for economic reasons. Several states have intervened to offer subsidies to prevent their plants from closing, essentially applying a Band-Aid while they figure out a long-term strategy. In late 2021, Congress finally came to the rescue with a $6 billion subsidy plan for the nuclear plants, which may be enough to keep them alive for the next few years, but it is not a permanent fix.

All of these issues have driven a broad consensus in the American nuclear industry that the country needs to move beyond twentieth-century nuclear designs to something new. The emerging view is that the industry needs to build fewer pieces of equipment on-site and more of the critical components in factories, where modern methods of manufacturing and quality control can be applied. Producing the major components in factories should, in theory, prevent runaway cost overruns; in fact, it ought to get cheaper the more often it is done. The nuclear industry, in other words, wants to tap the power of the learning curve. Thanks to advances in technology, this newer generation of plants would also be safer than the old ones—even if all the active safety systems in the plant were somehow knocked offline, the plant would still shut itself down without blowing up or leaking radiation. One technique would be to configure the nuclear fuel so that it cannot get hot enough to cause a reactor meltdown.

Many young engineers and entrepreneurs have thrown themselves into this work, starting new companies with bold plans, and Silicon Valley has tentatively proven willing to finance their ideas. Private efforts are even afoot to develop reactors fueled by nuclear fusion rather than fission—that is, plants that would create heat by fusing hydrogen into heavier elements, instead of splitting heavy elements into lighter ones, as today's reactors do. Fusion is the process that powers the sun, but harnessing it for use on Earth has been fiendishly complicated. A wave of recent announcements by fusion companies has raised hopes that they might finally be on the right track.

If a new generation of nuclear reactors is what we need, that still seems to be a long way off. Congress has taken some steps to modernize the nation's nuclear enterprise and its regulation of the industry, but for decades, did not see fit to put up the money for a serious development effort. Once again, that changed in late 2021 at the behest of the Biden administration, with the president's infrastructure bill committing more than $3 billion for advanced reactor demonstrations. It's just a down payment on what is really needed.

While it is possible that nuclear power will never make a major contribution to solving the climate challenge, it would be wonderful if that assessment were proven wrong. For that to happen, politicians must replace feel-good rhetoric about nuclear power with action. The nation needs to figure out, once and for all, if it can find a path out of the nuclear conundrum—or if it needs to give up and move on.

There are some bright spots. A company called NuScale Power has come a long way in designing a new type of modular reactor, meant to be built in factories and then hooked up on-site. It's not really one of the new-age designs we discussed above—it's more of a scaled-down version of the existing technology. But by going modular, the company may solve some of the existing construction problems. Large power plants could be made by stringing several modules together.

The company is getting support from the federal government and has settled on a site in Idaho for the first project. If the project ultimately goes well, we may know by the late 2020s whether the NuScale reactor works and can be built without massive cost overruns. If so, then NuScale might become a contributor to the energy transition, functioning as a bridge form of nuclear energy until the more modern designs come to fruition. In the meantime, the nation cannot assume that nuclear power will solve our problems. We need to move forward on other technologies.

Put It Back?

The politicians in Queensland, one of those Australian states with enormous coal mines, were buzzing with excitement. They had made up their minds to demonstrate, once and for all, that a potential solution for the emissions problem was at hand. The idea had been discussed for decades and, in principle, was simple as can be: just capture the emissions from power plants and factories and inject them into the ground. If Mother Nature has spent millions of years socking away carbon in underground reservoirs, in the form of coal, oil, and natural gas, couldn't human beings make use of that carbon and then stick the waste right back where it came from?

This idea was supported by the observation that every technology needed to make the scheme work had already been tried and proven at a commercial scale. In fact, oil companies, especially in the United States, were regularly pumping carbon dioxide underground as a way to increase the pressure in reservoirs, forcing out more oil. And the Norwegians had spent years proving that CO_2 pumped underground in the right spot would likely stay down. All you really needed to do, the optimists believed, was put all the pieces together.

In fact, elements of the fossil-fuel industry got so excited about this idea that they got ahead of themselves, bandying around the marketing slogan "clean coal" as though it had already been accomplished. But at the turn of the twenty-first century, huge questions still hung over the practicality of the approach, which came to be known as carbon capture and storage. With support from the government of Australia, the state of Queensland, and Australia's coal industry, a fellow named Chris Greig intended to answer the biggest of those questions: What would it really cost to build and operate a full-scale power plant that could burn coal with no emissions?

Rosy forecasts had been published claiming these projects could be built and operated for a modest premium over the current costs of electricity. But after spending the better part of a decade and $100 million creating studies and engineering plans for a power plant optimistically called ZeroGen, Dr. Greig and his colleagues determined that the Australian and Queensland governments would need to guarantee the plant an electricity price four times the market rate in order for it to deliver an acceptable return.

Some of the difficulty was just bad luck. Queensland had settled on a spot in its coal-producing region where the underground formations proved to be unsuitable for holding large amounts of carbon dioxide. They were simply too dense; an underground sandstone layer needs to be fairly permeable to accept huge amounts of CO_2. By the time all the test drilling was done and this became clear, the project was on shaky ground. In theory, ZeroGen could have built a pipeline to pump its CO_2 into a more suitable formation, but that would have added more costs to an already costly plan. Ultimately, the project was canceled.

In a broader sense, the ZeroGen project was emblematic of a worldwide problem as it gradually became clear that "clean coal," at the scale needed to make a dent in power emissions, was not going to be simple at all. Getting all the pieces of such a complex power plant to work together reliably was going to be a profound engineering challenge. The costs would surely come down by the time several plants had been built, but would they come down enough to make this kind of plant a competitive source of low-carbon power? The world still does not have an answer to that question.

ZeroGen was canceled in 2011, by which time the United States was pursuing a "clean coal" project of its own. Called FutureGen, it was to be located in Mattoon, Illinois. This one, too, was plagued by high costs and by irresolution on the part of the government about whether it would be willing to spend the necessary money. FutureGen was canceled, restarted, relocated, and canceled again, eating up $200 million before the final cancellation in 2015.

You would think the industries that depend on fossil fuels would have a huge motive to take the lead in developing carbon capture and storage technologies, and indeed, some companies have tried to get a few projects off the ground. But several of these projects have not fared much better than the big government-sponsored ones. Southern Company, the huge holding company whose subsidiaries supply much of the American South with electricity, led a project in Mississippi to try to prove the technology. But the company made a critical error: it spent only a fraction of what the Australians had on the basic engineering and design plans, and started construction well before those plans were complete. Costs eventually skyrocketed, the company had trouble getting its carbon-capturing technology to work, and it eventually gave up and started running the plant as a conventional power plant, fired by natural gas and with the emissions simply released to the atmosphere. In 2021, workers finally blew up the part of the plant that was supposed to capture carbon dioxide. Mississippi, one of the poorest states in the United States, will be saddled with some of the project's costs for decades.

After that sad track record, you might think most experts would be deeply skeptical about this whole approach to the emissions problem. But you would be only partially right. Carbon capture and storage still holds some promise, and, in fact, a new wave of projects has been launched recently. This technology may prove to be essential for solving certain kinds of emissions problems. If new methods of making cement, steel, and some chemicals do not pan out, then the only way to clean those industries up may be to capture the emissions and bury them. This means that while governments need to learn from the mistakes of the past, they also need to keep investing in the technology's future. As the United States ramps up its research on clean energy, a fair slice of the money needs to go into figuring out whether this technology has something to offer.

Some signs of hope have appeared on the horizon. Shell, the big oil company, is testing a carbon-capturing project at a site called Boundary Dam in Canada that, despite serious start-up problems, now appears to be operating reasonably well. A carbon-capturing plant has also begun operating in Texas. It is a pilot project, much smaller than a commercial plant would be, but it may work as a proof of concept. The plant is based on an exotic technology that yields a stream of pure carbon dioxide ready for burial, and could prove more affordable than other carbon-capturing power plants. Britain has announced plans for a larger plant of this type.

The United States Congress recently offered special help to companies working on carbon capture and storage. These companies can now take a tax credit of as much as $85 a ton for sticking emissions into the ground, a price high enough to make the economics of some projects more favorable. We may thus get some serious tests of the approach over the next decade. Yet if we are really going to continue using fossil fuels and simply capturing the emissions, the scale of the challenge ahead of us is immense. If we used the world's existing oil infrastructure—all the pipes, all the pumps, all the ships, all the rigs, and all the wells in the world—to stick carbon dioxide underground, that would handle less than 10 percent of the world's current emissions. And that does not count the equipment that would be required to capture the waste streams containing CO_2 that issue from plants and factories, refine it so that it is sufficiently pure, gather it from thousands of sources, and compress it so that it can be squeezed underground—all of which would cost money and use energy.

Most people working on carbon capture are well aware of all that. Yet the pursuit continues.

Hot Rocks, Everywhere

People were just getting off work that evening, dodging rain showers as they headed home or to shops to pick up Christmas gifts. Then the ground started shaking.

In the newsroom of the city's main newspaper, reporters unaccustomed to earthquakes dived under their desks—and had to be ordered by a senior editor to get outside and find out what the heck was going on. A young waitress named Aysel Mermer thought a bomb had gone off. A receptionist named Eveline Meyer, who was at home when the shaking started, guessed that her washing machine must have come to life on its own and started clattering with an unbalanced load. "Am I now completely nuts?" she asked a friend on the phone.

In reality, the elegant city of Basel, Switzerland, had just endured a 3.4 magnitude earthquake, fairly small as earthquakes go but big enough to cause fear and damage in a city whose last major quake had been 650 years ago. That earlier quake, in 1356, was a natural event. The one that struck Basel on December 8, 2006, decidedly was not.

It was caused by a drilling project taking place at the corner of Neuhaus Street and Shafer Lane, where a rig had managed to punch a hole in the Earth three miles deep. Years later, James Glanz, a reporter for the *New York Times* who reconstructed the day's events, described the project as an attempt to tap "a vast source of clean, renewable energy that seemed straight out of a Jules Verne novel: the heat simmering within the Earth's bedrock."

Just after the earthquake rocked the city, one of the executives of the drilling project was hauled downtown in a police car to explain what had happened. The project was hastily shut down. But it had achieved an unintended aim: helping to clarify one of the major risks that would be involved in trying to tap geothermal heat as a source of power.

This was an inauspicious beginning, to say the least, for this new concept in clean energy. But as readers of this book have learned by now, all sources of clean energy pose risks and difficulties, just as the fossil energy sources they are meant to replace do. The events in Basel may have slowed down the deployment of similar projects, but they did not kill interest in this approach. Instead, they helped to frame the right questions, particularly: Can the risks be managed well enough to turn geothermal energy into a major power source?

Volcanoes remind us, sometimes at terrible scale, that the interior of the Earth is exceptionally hot, despite its cool, habitable crust. In some places, this heat rises to a point near the surface, and in those areas, conventional technology can already be used to tap it and make electrical power.

Up to now, the technology has been limited to particularly favorable locations. Most American geothermal plants are in the West, which is more geologically active, with heat rising nearer to the surface, than in the East. California gets almost 6 percent of its electricity from geothermal plants, most of them in a region called The Geysers north of San Francisco. Generally, the plants use wells to inject water deep into the hot rocks, then use other wells to produce the resulting steam, which is used to turn a turbine attached to a generator. Some countries are especially favored with conventional geothermal potential; Iceland, for instance, gets almost all its electricity from geothermal heat. But you won't find geothermal plants anywhere near the East Coast of the United States, or in most other parts of the world—the hot rocks are too deep underground.

For a couple of decades now, scientists and engineers have realized they may be able to overcome this limitation—because if you drill down far enough, you can find hot rocks everywhere. Deep rocks are not likely in contact with groundwater, and may consist of nearly impermeable granite or other dense stone, so it may be much harder to tap them for power than in the older projects out West. In a major report in 2006, however, researchers at the Massachusetts Institute of Technology explained how these limitations might be overcome. Hydraulic fracturing—the very technology that has allowed energy companies to pull vast amounts of oil and gas out of previously impervious rock layers—could be turned to another purpose, the report found. You could drill down deep enough, fracture the rock, inject water, and recover steam capable of running a power plant. The report calculated, in fact, that this would be an immense source of power—enough in the United States, at least in theory, to satisfy the country's entire energy demand two thousand times over.

The biggest risk is probably the one identified in the Basel drilling project: "induced seismicity," which is a fancy way of saying the drilling and the associated hydrofracking may cause earthquakes. Even the older type of geothermal energy, operating at shallow depths, can cause earthquakes, though these are generally small. But as the Basel event showed, the quakes induced by deeper drilling can be big enough to throw a whole city into turmoil.

If geothermal energy is to become a far bigger power source, we need to know if that risk can be ameliorated. Is it possible to identify areas that are unlikely to generate earthquakes, and thus might be particularly suitable for drilling? Conversely, can especially high-risk areas be identified and excluded? Can alterations in the hydrofracking technology minimize the risks? One startup company, in Canada, believes it may be possible to build geothermal power plants anywhere in the world without hydrofracking at all—using drilling to install heat-collecting pipes deep underground.

Beyond safety issues, the biggest question about geothermal technology is the obvious one: How much is it going to cost? Geothermal wells require a different set of fracking techniques than those used to produce oil or gas. The cracks required for geothermal are much smaller. The working temperatures are higher. And the material—hard rock—is harder and tougher than the shale that holds oil deposits. Even if it is possible to build these geothermal plants, we do not know yet if it will be affordable. In the coming economic competition to supply flexible power to meet the needs of a grid with large amounts of wind and solar power, proponents of geothermal energy will need to prove that low-risk plants can be built at a tolerable cost.

The good news is that this is one ball the United States government has managed not to drop entirely. The Department of Energy has shown interest in the technology since the 2006 MIT report. It is funding early demonstration projects and, more recently, identified a site in Utah that will become a major test bed for improving drilling techniques and figuring out how to build these power plants. Even that is not enough, though. This promising technology deserves a much harder push than it is getting, with more federal money pumped into proving the basic concepts. Other countries also need to do far more—particularly China, which has an immense amount of geological work to do just to understand its rock formations and how they might be exploited for clean power. Electricity buyers can play a role, too. Google recently signed a deal to buy the output of an especially promising project under development by a company called Fervo Energy.

It is not entirely certain that this source of power will be needed, of course. But if advanced nuclear power fails, or if burying emissions from fossil power plants proves impractical, geothermal could well prove to be essential in the later stages of the energy transition. Countries—and companies—should be investing in it as though it were a potential world-saving technology, because it might ultimately prove to be one.

A Culture of Innovation

The technologies we have cited in this chapter deserve public support to develop, to test, and to see if they are practical. Most of them involve ways to inject clean power into the electric grid, consistent with the "electrify everything" approach to the emissions problem, which entails cleaning up the grid while moving away from fuel-burning equipment like gasoline cars or gas furnaces and powering those needs with electricity. Our list, then, is focused on the medium term: the technologies we need to push toward success or failure in the next decade. But these electrical technologies are not by any means the only kinds of innovation we need over the long term. Society has to learn to satisfy our whole suite of modern energy and material needs in the cleanest, most efficient ways possible.

Many exciting technologies are still in the earliest stages: under development in university laboratories or stuck somewhere on the rocky road from the laboratory to the marketplace. People are working on batteries, for instance, that might eventually carry three or four times as much power as today's and might recharge in minutes. They are working on more energy-efficient ways to construct buildings, on ways to heat and cool those buildings using miserly amounts of electricity, on protocols to get all our devices cooperating with the grid to use power when it is most available and thus cheapest. We will need exotic new compounds, new and cleaner ways of mining metals like lithium and cobalt, new ways of recycling batteries and solar panels and wind-turbine blades.

But we must remember that new materials and new machines are not the only kinds of innovation we need. They may not even be the most important. We also need innovation in markets, so that they properly value low-emissions technologies. We need innovation in public policies, so that society sets long-term emissions goals and writes them into law, giving everyone a clear sense of where we need to go. We need experimentation and innovation in ways to alter behavior, like the targeted tax on meat that we mentioned earlier in the book. Many conservatives, and many economists, may argue that we are talking about decisions that ought to be made in the markets, not by governments. And we agree, to a point: wherever markets can be used to discover the cheapest ways of meeting targets or goals, they ought to be. A competitive market is the best price-finding mechanism that human beings have ever created. But it certainly does not follow that *all* important economic—to say nothing of ethical—decisions can be made by the markets. The world is rife with problems that can only be solved by public standards and collective decision-making. The moral imperative to cut emissions enough that we preserve the livability of the planet for future generations is one example of a decision that cannot and will not be made by markets acting alone.

Mariana Mazzucato, a professor who studies innovation policy at University College London, argues that the supposed divide between "the market" and "the government" is false at heart. Many technological innovations that we now take for granted can trace their origins back to government money, or to government as an eager first buyer. The most famous example is the Internet, which was launched by a Pentagon unit that funds innovative, off-the-wall ideas. But a slew of technologies critical to the smart phone also started this way, including touch screens, cellular data, and GPS. In fact, the entire computer revolution grew from early Pentagon contracts in the 1950s, in which the government paid high prices for newly invented semiconductor chips. The pioneers of Silicon Valley survived for years on this government largesse.

In the realm of energy, federal research, grants, and product purchases have stimulated advanced turbines, used to generate power from natural gas. They pioneered the 3D seismic imaging and the fracking technologies that have resulted in America's oil and natural gas boom. They paid directly for the huge Bonneville Power system in the Pacific Northwest, for rural electrification, and for the development of nuclear power. One must struggle to find developments in big energy technology that did not get a helping hand from the federal government.

To forestall the looming climate crisis, both governments and corporations must focus much more intently on the new technologies we will need to displace the fossil energy system—not just on creating these clean technologies, but on pulling them out of the laboratory and getting them up to scale. This will mean pumping billions into demonstration projects and first-of-a-kind power plants. It also means walking away from favoritism toward particular technologies and focusing on the real goal: cutting emissions.

The issue is not just what technologies we need for the energy transition, but whether we can roll them out at a pace and scale fast enough to head off the worst damages. This question is particularly acute in the United States, where much of the country's public infrastructure—power plants, highways, water systems, and so on—was built in the middle decades of the twentieth century and is now aging fast. In recent years, the political culture in the United States has become so toxic that the government cannot even raise the money to maintain these old systems, much less build new ones. In the early days of his presidency, Joe Biden called for a huge wave of public investment in updating the country's infrastructure, and Congress passed a bold first step. The issue in America is whether the country can bridge its political divide and once again learn how to tackle big national projects.

Corporations have a critical role as well. They are already playing one, to a degree. Hundreds of American corporations have set targets for reducing their own emissions of greenhouse gases, with many of them going into the marketplace and buying renewable energy directly from wind or solar farms. But that number needs to become thousands. Financiers have begun to realign their investments, moving away from fossil fuels and toward clean energy. But many of the blockages that are restricting investment in solving the climate crisis, including weak public policy and shortsightedness among companies, need to be blown apart.

The task sounds daunting, we know. So how can ordinary people help to steer global finance toward a cleaner economy?

Pulling the Levers

One way, we think, is the method we already mentioned in Chapter 7. People working inside corporations need to speak up about the climate problem, and ask questions about what their own company is doing about it. Some of these questions should be quite basic: Has the company taken responsibility for its emissions? Is it working to cut them? How? Is it taking responsibility for emissions throughout its supply chain? If you work in a large, publicly traded corporation, ask whether it is publishing an annual sustainability report, and whether it has been meeting its own targets as laid out in that report. Are those targets consistent with the goals of the Paris Agreement?

If you work in a company that makes technology investments, or finances the deployment of infrastructure, you have an even bigger opportunity. Is the company looking at cutting-edge technologies that promise to help cut emissions, like the ones we outlined above? Is it taking investment positions in any of them? Is it putting enough capital at risk to show that the company is serious about its climate commitment? Too many companies with multi-billion-dollar investment budgets are spending $10 million here or $20 million there to dabble in green technologies. This is not a serious program; it is greenwashing. In fact, some of these minor investments are even paid for out of corporate public-relations budgets.

Delicate as it may be to confront the top executives of your corporation about their climate commitments, they need to be confronted. Young job-seekers who are considering joining a company can ask these questions, too. If you have signed the Climate Voice pledge that we mentioned in Chapter 7—and you should—then you have made a promise to raise these issues before going to work at a company. Ask the questions. And if you get mealymouthed answers, go to work somewhere else and tell the company you are ditching exactly why.

Ordinary people can influence corporate behavior in another way, too. If you are among the lucky half of Americans with an investment portfolio, take a hard look at what stocks you own, and whether they are truly aligned with your environmental values. This can be tricky to do if your holdings are largely in mutual funds, since they may be actively managed, with the exact portfolio of stocks changing daily. Fortunately, funds that take environmental and social goals into account are proliferating on Wall Street. The managers of these funds have pledged to their investors that they will avoid certain kinds of stocks—oil and coal companies, for instance. The buzzword on Wall Street for this sort of investing is "ESG" funds, with the letters standing for the idea that environmental, social, and good-governance criteria are applied to screen investments. We know that some people avoid actively managed mutual funds, given their high fees, but increasingly, you can also invest in passive index funds that track a broad selection of stocks with favorable environmental ratings. The big investment houses, like Fidelity and Vanguard, all offer such financial products now. The more investors like you take environmental criteria into account, the greater the effect will be, influencing the financial decisions that companies make—and thus helping to speed new technologies down the learning curve.

This is not the only reason to clean up your investment portfolio, though. A decade ago, college students across the country asked their universities to take a stand on climate change by divesting themselves from fossil fuels. Some small schools led the way, but many big names—including Harvard, Yale, and Stanford—refused for years to divest. Over the succeeding decade, the stock prices of many fossil-fuel companies tanked. As we mentioned earlier in the book, all of the large American coal companies went through bankruptcy, wiping out their stock investors, but even mighty oil companies like ExxonMobil saw their stock prices swoon amid low oil prices. University endowments refuse to publicize their investments, so we cannot say for sure, but it is quite likely that some of the big endowment funds lost money on their fossil holdings. If they had done what the students asked them to do in 2012, they would have saved themselves that pain. In 2020 and 2021, the schools we mentioned above finally, and sheepishly, committed to get out of fossil fuels.

Don't be like those universities. You want to move your money onto greener pastures not just because it is the right thing to do, which it is, but because owning a piece of the fossil economy is an active risk to your portfolio. If the world does get serious about solving climate change, it means the oil companies will be sitting on billions of barrels they cannot produce, and the coal companies on megatons of black rocks they cannot dig up to burn.

The term of art here is "stranded assets"; these assets now help to inflate the stock valuations of the companies, but they will—we hope and expect—be useless in the long run. In thinking about your own money, put your poker chips on the future, not the past. The writer Bill McKibben, who helped to jump-start the divestment movement, offers a simple rule of thumb for investors.

"If it's wrong to wreck the planet," he wrote in 2018, "it's wrong to profit from the wreckage."

CHAPTER 9

Saying Yes

"Got our breakfast as early as usual, and ready for a start by six o'clock," a man named James Abbey wrote in his diary on the morning of April 30, 1850, as his family was trudging across the territory that is now Wyoming. "The day is cold and the wind is blowing so hard that it is almost impossible to stand up, but the boys say we are bound for California and it will never do to stop for wind, so we toddled on."

The Abbeys were headed west to take their chances in the California gold fields, discovered only two years earlier. Settlers like them who wanted to cross the Rocky Mountains with cattle and a caravan of covered wagons found there was only one good place to do it. The passage is known as South Pass, in southern Wyoming, with an elevation topping out around 7,500 feet. The approach was so gentle that wagons navigated it with relative ease. As many as half a million settlers poured through that pass between 1840 and 1860, seizing the western half of the continent for the growing American nation.

As they approached the pass, every single one of them noticed that southern Wyoming was a place of powerful winds. "The wind blew hard enough all night to make a fellow almost believe himself to be at sea in a gale," Mr. Abbey wrote a week after the diary entry above.

The explanation for the stiff winds of southern Wyoming is the very geography that drew the settlers toward that mountain gap. South Pass is part of a broader dip in the towering Rockies, a low saddle twenty-five miles wide or so. The Shoshone tribe of Native Americans reputedly called it "the place where God ran out of mountains."

Geographically, that break in the cordillera functions much like the California mountain passes we described earlier, but on a larger scale. It helps to concentrate the winds flowing from west to east. The air flow meets little resistance on the nearly treeless landscape, helping to turn southern Wyoming into one of the windiest places in the United States. A broad band across the southern part of the state is known today as the Wyoming Wind Corridor.

Wyoming settlers in the latter decades of the nineteenth century filled their own diaries with complaints. In the winter, the wind could pile snow fifty feet high, destroying herds of cattle or sheep and burying houses. Some of the settlers, the ones who read a little geology, understood that the bitter wind had helped scour away the topsoil, leaving Wyoming a poor place to farm. Through hard trial and error, they discovered that the most economic use of the land was to graze cattle and sheep, but the grass was so thin it took a lot of land per animal; a large Wyoming cattle ranch can cover thousands of acres. The modern economy of the state is founded on ranching, mining coal, and extracting oil and gas, with a significant boost from tourism. The early settlers used small windmills to pump water, but through most of the history of the state, nobody would have imagined the wind as one of Wyoming's most valuable resources. Even today, it is more often seen as an affliction. On a single day in 2017, wind gusts toppled fourteen tractor-trailer trucks on Interstate 25 in northern Colorado and southern Wyoming. A twenty-mile stretch of the highway had to be closed.

At the wheel of a smaller truck—a black pickup—not long ago, a fellow named Bill Miller explained why he had come to see Wyoming's winds not as a curse, but an opportunity. He was driving across the dirt roads of a ranch, a place called the Overland Trail Cattle Company, one of those vast landholdings that seems possible only in the American West—five hundred square miles of rugged rangeland. The ranch, as it happens, is due east of that large wind funnel in the mountains, putting it right in the middle of the Wyoming Wind Corridor. At the top of a hill, Mr. Miller stepped out of his truck, and the wind whipped up tufts of his white hair and his white goatee. Mr. Miller swept his arm across the horizon, taking in the immensity of the ranch, the rugged hills, and the broad, arid valleys stretching as far as the eye could see. As cattle ambled across the landscape, black specks in the distance, he sketched a vision of the future. For the Overland Trail Cattle Company is not going to be a simple Wyoming cattle ranch for much longer.

Mr. Miller is no simple rancher, either, even if he does manage the ranch. He is an oil man as much as anything else—the right-hand man for one of America's billionaires, the Denver oil and real-estate baron Philip Anschutz. The Anschutz fortune—some of which goes to support conservative causes—is built on natural resources: land, oil, and soon perhaps, wind. On the ranch, Mr. Anschutz and Mr. Miller are planning to build the largest wind farm in North America, close to a thousand turbines capturing those powerful gusts emerging from the mountains. They want to send the power through a special high-voltage line across a huge stretch of desert to the largest electricity market in the western United States, the 40 million people of California.

Like his boss, Mr. Miller leans right, politically. Bouncing across the dirt roads of the ranch, he took pains to point out that he is no fan of government programs like the Green New Deal, a proposal to solve America's energy problems through a vast amount of federal spending. "Hell no," Mr. Miller said as he steered the Ford F-150 Raptor across a hundred miles of dirt roads one day in 2019. "I just don't think that makes sense. But it really doesn't matter what I believe. The policymakers and the voters and the people in this country are going in this direction, toward clean energy. If we're going to be a viable natural-resource industry, we had better stand up and help them out."

Mr. Miller and Mr. Anschutz decided to build a wind farm almost fifteen years ago. And yet, as we completed this book, they still had not erected a single turbine. That long delay stemmed in part from a specific federal law, the National Environmental Policy Act. The law came into play because the Anschutz Corporation does not own the entirety of the ranch it runs; it owns about half the land, and the federal government owns the rest. Mr. Miller, like ranchers across the American West, leases grazing rights for the government parcels, which are intermixed with the private land in a checkerboard pattern. As you cross the ranch, nothing indicates whether you are on private land or government land. But legally, it matters a great deal.

When Mr. Miller first broached the idea of a wind farm, he knew he would need special permission to build on the federal portions. Somebody warned him that it might take five years to get the necessary permits. "You have got to be crazy," he remembers thinking. His optimism was wildly misplaced; it has taken a dozen years and counting to get most of the permits and legal documents he needs to move forward.

The National Environmental Policy Act, or NEPA, was passed in 1969, decreeing that if the federal government was going to take any action with a potentially significant effect on the environment, it had to conduct a study of the possible consequences, including ways to minimize damage. In the decades since, the law has become an immense hurdle for large new infrastructure projects. Passed as an environmental protection measure, it is now having a perverse effect: it sometimes stops or slows clean-energy projects whose purpose is to save the planet from environmental ruin.

In the early years of NEPA, the required analyses were sometimes twenty or thirty pages long. Nowadays, they can run tens of thousands of pages. Complying with the law, and with the demands that multiple federal agencies make in their efforts to obey it, can require the hiring of biologists, engineers, hydrologists, soil scientists, and many others. Much of this work is defensive, because people opposed to projects for any reason can and do file lawsuits to stop them. Federal agencies stockpile reams of paperwork to gird themselves for these legal battles. Not only did the Wyoming wind farm require a massive effort, so did the power lines that will carry electricity 732 miles across three states to reach an electricity hub in Nevada that is linked to California. Mr. Miller estimates that the Anschutz Corporation has spent close to $200 million just to accumulate the towering pile of paper decreeing that it can start building. And the matter is not fully resolved, either: as this book was going to press, a conservation easement meant to protect sage grouse was holding up a segment of the required power line. That fight is about thirty acres of right-of-way on a sixteen-thousand-acre ranch, but it might be enough to kill the Anschutz project.

Lengthy delays like this are injurious—not just in money wasted, but in stalling the rapid energy transition we need. If America is to meet the goals of the Paris Agreement, it must rapidly accelerate the construction of wind and solar farms, to a pace at least three or four times as fast as that achieved in recent years. All that wind and solar power will have to be moved around the country on a network of high-efficiency power lines much denser than the one we have today. To stay on track, we have less than thirty years to clean up the entire energy system of the United States. It is simply not going to happen if every major project needed to achieve the goal is going to be held up by ten or fifteen years of paperwork.

Mr. Miller has started construction on the roads for the wind farm, but turbine installation is not likely to proceed until he has signed deals to sell the power in California, or possibly elsewhere in the Southwest. Those deals may not come until it is a legal certainty that he can build. He remains optimistic.

"We will not fail," Mr. Miller said as he drove across the ranch. "The difficult we do today; the impossible will take us a little longer."

Grand Ambitions

In this book, we have argued that a great deal of climate action has to occur at state and local levels in the United States. State governments have power over the electric grid, and local governments control the zoning rules that determine how leaky new buildings are. We think it is particularly important to keep working at the local and state levels during periods when Washington is paralyzed, as it so often seems to be lately. State governments are often called the laboratories of democracy; they are much closer to the people than the national government, which means citizens can use their voices to move climate policy in the right direction. Adam Browning, who used to head an advocacy group called Vote Solar, had a motto for his staff: "If your plan involves Congress, it's a bad plan."

Yet, while we firmly believe that local action matters a great deal, let us also be clear about this: America needs a national climate law. It was a tragedy in 2009 that an ambitious climate bill failed for lack of a handful of votes in the Senate. Since then, through three presidential administrations, no other bill of such scope has come close to passage. The two measures that Joe Biden pushed through in 2021 and 2022 were certainly steps in the right direction, but they were not as broad as that failed 2009 bill. Some problems can only be solved by Congress. The thicket of regulations that holds up construction of new power lines that are clearly in the national interest—only Congress can fix that. The misuse of federal environmental laws to stop projects that are clearly beneficial to the environment—Congress needs to fix that, too.

If the planet is to be saved from the worst ravages of climate change, here is how we think it has to happen: A rising political demand in the United States must prompt the government, at all levels, to move much faster on the energy transition. The policies that we have outlined in this book—such as ambitious targets to clean up the grid, far tighter building codes, and Buy Clean policies for major commodities like steel and cement—must be put into place over the coming decade. Other policies already in existence, like the state clean-electricity standards we mentioned in Chapter 2, need to be strengthened. The weak pledges and promises so common across the American landscape today, like the Energy Star rating system for new appliances, need to be turned into mandatory standards that drive mediocre products out of the market. As these policies take hold, the pace of the emissions decline in the United States will begin to accelerate.

We think other countries would react to this changed American environment by speeding up their own energy transitions. Some of them have, for decades, been much more committed to solving the problem than the United States has been, but the size of the American economy dwarfs that of these early climate champions, like Germany and the United Kingdom. The United States has the size and scale to help bring innovative energy and transport systems into existence. That does not mean the United States can do this alone; we need engineering help from the Germans, creative policy ideas from the British, manufacturing prowess from the Chinese, and so forth. But we do not think the global energy transition will happen as fast it needs to until the United States fully commits.

About twenty countries are responsible for the bulk of the greenhouse emissions pouring out of the world's factories, power plants, and tailpipes. A push in those countries would help to pull the rest of the world along. The two biggest emitters among those twenty are the United States and China, whose economies are highly intertwined. Despite their differences over human rights, trade, and other issues, it is critical for the fate of the planet that the two countries work together to set common emissions goals and to lead the world in developing low-carbon technologies.

India, the world's second most populous country, is already the world's number three emitter of greenhouse gases, even though tens of millions of people there still lack access to electricity. Across

Southern Asia, Africa, and Latin America, nearly 800 million people still live with no power, and for many millions more, the local grid is unreliable. Solving this problem has become an official goal of the United Nations, which means we are at an inflection point. These poor countries could, with moral justification, take the high-emissions route that Western countries took, burning vast quantities of coal and oil to power their economies out of poverty. The West has promised to help them figure out how to do something else: to skip the dirty phase and make a leap, going straight from too little energy to abundant clean energy.

The only way this will work is for rich countries to go first in scaling up the clean-energy solutions the world needs. Remember the magic of the learning curve: what matters for a new technology is getting to scale, which requires "overpaying" when a technology is still immature. Rich countries can afford to do that; poor countries cannot. If we choose to do this, technologies that look too costly today will be cheaper tomorrow. We have seen this in everything from Ford's automotive assembly line to electronics. In electricity production, it is already happening; solar and wind are now the cheapest new energy sources in many places, and the effects are filtering down to the poorest parts of the world. The Indians have been cutting back their plans for coal-burning power plants and ordering up immense new solar farms instead. Some of the cheapest solar prices in the world have been seen on large projects under construction in India. India has also embraced the LED lighting revolution and has pushed billions of bulbs into the market, cutting the projected growth in electrical demand on the country's grid. On a smaller scale, the clean-power revolution is occurring in Africa, too: tens of millions of people now have light at night via simple battery-powered lanterns charged from small solar panels.

But as we've already discussed, the innovations that have occurred will not be enough. In heavy industry, like steel and cement production, change has barely begun. Electric vehicles are getting cheaper, but they have not won the market yet. Construction standards across much of the world are still appalling. Bad urban design risks creating high-carbon, car-dependent, unhealthy cities housing billions of people.

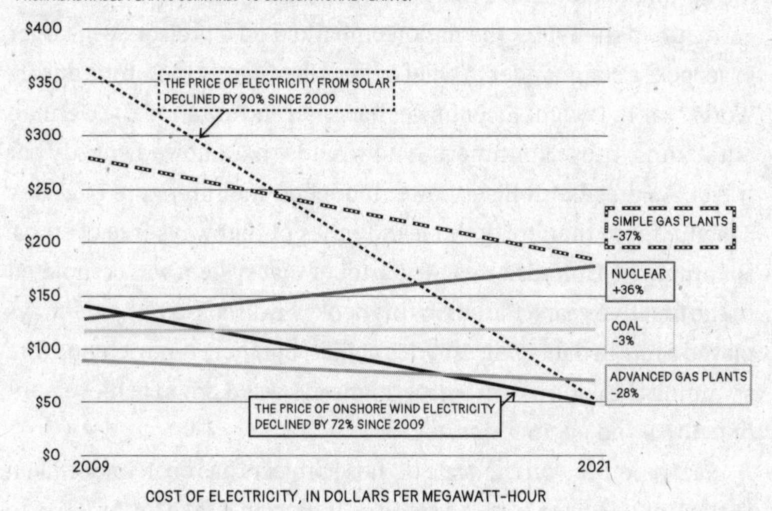

The American goal in tackling the climate problem has to be threefold: to clean up the existing dirty energy economy at home, to scale the market for emerging clean technologies so that they are soon affordable everywhere in the world, and to lead the way in smart design, setting an example for other countries to follow. Setting emissions targets that would not have to be met for decades, as so many politicians and business leaders have been doing, is not enough. Now, not later, the nation has to embark on a crash program to cut its own emissions. Because if we don't start now, it's going to be impossible to reach zero by 2050.

Committing to slashing our emissions would be nothing less than a grand national project. We know that such projects are possible because the United States has pulled them off before. In the 1930s, few Americans living in the countryside had electricity. Farm families were still hauling water in buckets and burning kerosene lanterns at night. Under the leadership of President Franklin D. Roosevelt, the nation embarked on a program to bring electricity to all Americans. The feat was largely accomplished in a decade, through federal loans to cooperatives organized by the farmers. Crews fanned out across the nation, hooking up a million farmsteads.

Again in the 1950s, the nation embarked on a project even larger in scope. As commander in chief of the Allied forces in Europe during World War II, Dwight Eisenhower had seen and admired the German autobahns, those shimmering new roads that allowed speedy car travel. As president, Eisenhower launched the Interstate Highway System, more than forty thousand miles of highways free of stops, spanning the United States. A quarter of that system was completed in the first five years. Earlier in this book, we discussed how freeways played a role in damaging America's cities, but there is no denying that spanning the country with superhighways was a grand achievement of politics and engineering.

Sadly, nothing on the scale of those projects has been undertaken in the United States in half a century. What will it take for America to tackle the climate problem at the speed and scale it requires?

Hard Choices

You might think money—turning on the spigot of federal largesse—would be the key to launching a clean-energy bonanza. That is the premise of the big bills that Joe Biden got passed early in his presidential term. But money is only a partial solution, and the gusher of federal dollars that Congress has appropriated must be spent judiciously. Repeatedly in the past, generous federal programs with lax controls have been magnets for fraud artists. We certainly think new spending programs are required, but they must be designed carefully, with rigorous controls to protect taxpayers' money.

Yet money is not the only problem. The National Environmental Policy Act, the law that tripped up the wind farm project in Wyoming, has long been recognized as a major issue. The administrations of George W. Bush, Barack Obama, and Donald Trump all tried to speed up the law and cut through red tape, but no president has the unilateral power to rewrite the law. This is a problem Congress really needs to fix. We do not believe the fix requires reducing or abandoning strong standards of environmental review. Instead, we envision setting strict time limits for the reviews, as well as shorter time windows for opponents to file lawsuits, so that all reasonable issues can be hashed out in two to three years. We know this kind of reform can be achieved, because other countries have passed laws to speed up their environmental reviews.

The federal government owns about 28 percent of the nation's land, with those holdings concentrated in the West. As a result, in that region, renewable-energy projects often run afoul of the National Environmental Policy Act. This is less true in the East; reasonably sized solar projects on private land in the Southeast often go through quickly. But even in the East, governmental red tape and legal hurdles can hold projects up. So can opponents who object to having to look at wind turbines or solar farms, and who use every twist of state or local planning laws to fight them. Putting wind farms up on land in the densely populated Northeast has become exceedingly difficult—one of the factors that is driving states there to order up huge offshore wind parks. But even with these offshore projects, fishing and boating interests have to be appeased.

We want to be clear about something: not every renewable energy project proposed in America should go through. Some pose genuine threats to special creatures, like desert tortoises or golden eagles. Some are just in the wrong place, and approving them would wreck scenic vistas, degrade cultural landmarks, or destroy important wildlife corridors. Without exception, renewable-energy projects are cleaner than the dirty coal or gas plants they're replacing, but they do generally cover a larger area, leading to the industrialization of the landscape. We understand why some people would object, and those people need to be heard. But the reality is that we are going to have to make many difficult decisions in the national interest. Projects should be steered to the most suitable locations, but they will need to go up somewhere.

Fortunately, renewable-energy development has been welcomed by local people across broad swaths of the United States—especially the Great Plains states, where powerful winds have made turbines an economic winner. In some counties in Oklahoma and Kansas, a significant fraction of local tax money now comes from wind farms, and those taxes help to educate the local schoolchildren. North Carolina counties have also seen an influx of tax money from the big solar buildout in that state. Many family farms have been put on a sounder footing by leases from wind developers, whose projects do not stop the farmers from growing their crops. Alas, many of the states where wind power is booming are among the least populous in the United States, so the power needs to be moved from there to big cities—one of the key issues for the success of the energy transition. To move electricity you need power lines, which brings up another thorny problem.

There may be no more difficult piece of infrastructure to build in the United States today than a long-distance power line. New ones are fought savagely at every step—by nearby landowners, as you might expect, given how unsightly power lines can be, but also by parties distant from the visual effects. States fight them, too, especially if the lines are going to carry power across a state without providing much local benefit. All these objections can stymie the lines no matter how badly they might be needed, in part because the federal agency that supervises the interstate trade of electricity, the Federal Energy Regulatory Commission, does not have the legal power to override state parochialism and approve lines of national importance. It can do that with natural gas pipelines, but Congress has not granted it the authority to do the same with power lines. In principle, power lines can be buried, just as oil pipelines usually are, but that drives costs up as much as fourfold, so underground power lines are hard to finance.

To get more than half the nation's power from renewable energy, the United States is likely to need a substantial network of new, high-capacity power lines spanning the country. China is building thousands of miles of these power lines; the United States has only a handful of them. Private developers have tried to jump-start that grand vision, and several have already failed, undone by political fights. Again, it is likely that only Congress can solve this problem, making clear that a modernized power grid is a top national priority and granting federal agencies the authority and the mandate to make it happen. A provision that passed Congress in 2021 may help to achieve that, but it was tucked into a spending bill, and we suspect a broader law will ultimately be needed. Perhaps recalcitrant states can be lured into supporting new power lines if they are promised some of the benefits, including cheaper electricity. Senators from the states in the middle of the country, the ones with all the wind, ought to be leading the effort to cut such a deal: harvesting their wind and exporting it to coastal cities would pour hundreds of billions of dollars into rural areas desperate for jobs and development.

The single most important thing Congress can do is to set an explicit national goal of cutting greenhouse emissions to net zero by the middle of this century. The phrase "net zero" means leaving room for some residual emissions, like those from planes or cement production, with those emissions fully offset by land uptake or other projects that reabsorb carbon dioxide. But the emissions allowance in 2050 will be exceedingly small; in practical terms, a net-zero goal means that major emissions sources like power plants and factories will have to figure out how to eliminate their emissions entirely.

In electricity production, Congress needs to set a national clean-electricity standard—that is, a limit on emissions that declines through time until the electricity system reaches zero emissions between 2035 and 2040. About 40 percent of the nation's power already comes from low-emission sources, including dams, nuclear plants, and wind and solar farms. The coal and gas plants supplying the rest need to be replaced. For the next decade, the main thing we must do is build more renewable power plants, chiefly wind and solar farms. Politically, a fivefold acceleration in the current rate of construction is a tall order. But physically and financially, that pace is well within the nation's capabilities.

Demanding Action

The election of Joe Biden in 2020 was a milestone in America's climate debate. For the first time in American history, a presidential campaign featured a robust discussion of the climate crisis and detailed plans from the winning candidate on what to do about it. It is an encouraging development, signaling that climate change has become a voting issue for a significant fraction of the American public. And yet Mr. Biden quickly faced political difficulties in getting his plans through Congress; the bills that ultimately passed were not as ambitious as what Mr. Biden had promised in the campaign. We hope, of course, that Mr. Biden succeeds in winning as much as he can on climate change. But the work we all need to accomplish will not end with the passage of the two climate laws that Mr. Biden signed in 2021 and 2022, nor will it end within a single or double presidential term. And while it will be enormously beneficial when Congress finally commits the full power of the federal government to solving our dirty energy problem, much of the detailed work still has to involve policies and lawmakers far removed from Washington.

The strategy we need to tackle the climate crisis has two prongs: pushing the low-emissions technology we already have as hard as possible, to displace emissions wherever possible as quickly as possible, and accelerating innovation for the parts of our economy where we do not yet have low-emissions solutions. Framed this way, it sounds easy; the reality, of course, is that the resistance of fossil-fuel industries and far-right ideologues, along with the general inertia of our political system and economy, have made both aspects of this strategy hard.

How do we accelerate action on both fronts? We know many people are weary of American political dysfunction, but there's just no other way: the answer lies in a more effective politics of climate change. It has been inspiring to see people—especially young people—marching in the streets all across the world to try to secure their futures on our overheating planet. Now we think the movement needs to grow more sophisticated in its demands, focusing its political power on specific targets. Calls for action must land where they will drive the particular policy changes that will deliver the most good. We cannot emphasize this enough: devote your political energies to the specific decisions that will drive change in those parts of the economy where emissions are the highest.

Over the previous eight chapters, we showed you many of the hidden levers that determine how our political and economic systems operate, most of them under government control. This means we all need to push our representatives on the issue—not just national politicians, but state and local ones, too. Vague campaign platitudes need to be met with citizen demands for specific plans and firm timetables.

Below, we'll recap some of the steps we need to take in each sector of the economy, but through a new lens this time: the lens of political leverage. For any given problem, what citizen actions at the town or city level can contribute to solving it? What about the county level, and then the state level? What role does the federal government need to play? And if it refuses, can states or cities find a way to work around it?

The most immediate opportunity in front of us is cleaning up the electrical grid. That entails pushing four existing technologies—wind turbines, solar panels, super-sized batteries, and digital management of electrical loads—to a much larger scale. It also involves holding on to aging nuclear power stations for as long as we can *safely* do so. For a decade or two, we will need to keep many of the existing power plants fired by natural gas but use them less and less, as renewable power supplants their production; construction of new gas power plants and pipelines needs to stop now. Remember that the American electricity grid is about 40 percent clean already. Existing renewable-energy technologies can probably get us to 70 or 80 percent clean at a reasonable cost. And by the time we get to that point, we will have invented more options to carry us the rest of the way.

How fast we go in adopting these technologies is largely under the control of state governments. The main control lever is in the hands of the agencies we talked about in Chapter 2, the public utility or public service commissions, which hold regular hearings regarding the plans of all the major utilities in each state. Citizens are invited to attend these meetings and given opportunities to speak. But in many states, very few people actually go. This is a shame, because utility commissioners have told us that when ordinary citizens do approach the microphone, they listen closely. For anyone fighting for a better climate future, it is imperative to align with advocacy organizations already working with utility commissions, and to turn out speakers anytime critical decisions are being made. These commissions need to push utilities to compare the cost of clean energy to the cost of running existing fossil-fuel plants, or of building new ones. Fortunately, this work is getting easier now that clean energy has become one of the most affordable ways to meet power demand.

These commissions take their basic marching orders from state legislatures, which means clean-energy plans also need to become a voting issue in state legislative elections. If your state has not set a goal to clean up its energy economy by 2050, start writing or calling your legislators to demand that they do so.

While we can go much faster on renewable energy, completing the transition to zero-emissions electricity will probably require new power sources that do not exist today, with the big rollout most likely occurring in the 2040s. These possibilities include geothermal plants—which might be built anywhere if new, deeper drilling methods are perfected—and new types of nuclear power stations that are cheaper, safer, and faster to build than the ones in operation today. Power plants that capture their carbon dioxide and store it underground are a possibility, too. Another exciting option is to convert clean electricity to hydrogen and store it for later conversion back into electricity. None of these options makes economic sense today, but, with serious research and development, some will one day. Funding this innovation agenda is largely the province of the federal government, and the measures Congress adopted in 2021 are a start, but states can help by doing some of the legwork. In particular, they should identify sites that might be appropriate for geothermal development, and they should start doing geological research on potential places for storing carbon dioxide underground.

Since renewable energy on the scale we need is going to require a lot of land, we have to start planning for environmentally zoned development. Governors can help by launching careful, statewide planning reviews, with the goal of designating areas that are appropriate for energy development while blocking off other areas as unsuitable. Local communities need to be approached early, and the potential tax revenues for their schools and other local services need to be highlighted. Towns that will suffer job loss and tax declines as fossil-fueled power plants are shut down will need special help, and ought to be at the top of the list for new renewables development. We believe Congress needs to come up with a national compensation scheme for communities dependent on fossil fuels, but until it does, governors and legislatures can do a lot to pursue fairness for them.

Getting the electricity system moving faster toward a clean future is the most urgent political target, but it is not the only one. As we outlined in Chapter 3, local building codes are a critical factor in determining whether our new homes and offices will be climate liabilities or climate assets. In some states, building codes are a statewide responsibility, so that strengthening them has to become a political goal at the state level. But that is not true across much of the United States; most localities decide what codes are going to apply within their boundaries. We cannot emphasize enough what a massive opportunity this is. If the local building code needs updating, a hearty and informed band of citizens going down to the city or county commission for several meetings in a row can probably get it done. Local builders may well resist; many oppose any demand that they spend a dollar now so that a homeowner can save $50 over time. But some “green builders” have become critical voices in alliance with citizens’ groups. As a minimum target, people need to demand that local building codes be updated to match the latest model codes from an organization called the International Codes Council. If your town is a little more forward-looking, though, demand the adoption of a “reach code” or “stretch code” with tighter standards.

You will recall our discussion in Chapter 3 that the use of natural gas in buildings needs to be phased out sooner rather than later. An update of the local building code may be a major opportunity to do this; even if codes do not ban gas outright, they can be structured in a way that encourages all-electric buildings and discourages the installation of gas lines. A related political target is to go after the subsidies that gas companies pay to get property developers to put gas appliances into new homes—effectively, legalized bribes meant to lock in gas use for decades. These programs are usually under the control of a state’s public utility commission. They need to be stopped, which will tilt the economics for developers in favor of all-electric buildings.

Another—politically tricky—set of targets needs to focus on energy use in *existing* buildings. As far as energy waste, decades-old buildings are a much bigger problem than buildings from the last twenty years or so. In Chapter 3, we discussed fire codes as a historical precedent and model for what needs to happen here: city governments not only enforce fire rules when a building goes up, but come around regularly and inspect larger buildings for compliance with those codes. Every city government in the country needs to make a mental shift, toward seeing energy waste as a problem on a par with unabated fire hazards.

Initially, cities should simply require data-keeping and reporting to the city government of energy use in buildings, with the data made public for commercial and industrial buildings. A couple of years after this law passes, the screws need to tighten: the city should require annual declines in energy use in each building, on the order of 5 percent a year, and it should fine building owners who fail to meet the targets. This proposal may sound radical, but several cities, including New York and Washington, D.C., have already done it; the trend needs to spread across the country. Many cities may hesitate to apply this rule to houses or small apartment buildings, so it makes sense to start with larger buildings, work out the kinks, and continue from there. The effect of this policy will be to combat the poor design and operation of heating and air-conditioning systems, which is common across the country. With the screws tightening every year, owners will have no choice but to get their systems in order.

In much of the country, people looking for a new home pay far too little attention to the energy use of the houses they're considering. This is somewhat understandable: except in a handful of locales, Realtors have fought off any requirement that utility bills be disclosed to potential buyers. This is crazy! Every city government needs to adopt rules requiring that basic utility bills be disclosed in every realty listing and sales contract. Ideally Congress would tackle this issue with national standards regarding energy disclosure, but until Congress acts, states and localities need to move forward. Many millions of buildings in the United States need to be retrofitted with better insulation in the walls, foam sealants to plug air gaps, and updated appliances. A sweeping national program offering cash grants or tax credits for such retrofits would be ideal. Congress put about $5 billion for such programs into the infrastructure bill it passed in 2021, but much more is needed, and states need to push this issue hard.

Utilities already offer rebates to customers installing new appliances; those programs need to be expanded and made stricter. For instance, the rebates should be paid only if the customer installs equipment of the highest efficiency rating, which would typically mean heat pumps able to respond to signals from the grid. Gas appliances should become ineligible for rebates. In fact, both gas and inefficient electric appliances ought to be subject to a surtax, with the money used to buy down the retail cost of the most efficient appliances. This will probably have to be done at the state level, but local governments need to be creative in finding ways to make heat pumps the market standard as quickly as possible.

As you'll remember from Chapter 4, three American states have declared a goal of banning the sale of gasoline-burning cars by 2035, emulating several European countries that have set similar goals. Governors of other states need to be pushed to follow their lead. At the same time, both states and cities need to work much harder to install the charging stations that will be needed as more and more electric cars hit the roads. One possibility is to order local utilities to do the work, and charge their ratepayers for it; that is what Colorado did in 2019. One reason carmakers are embracing electric cars is that in fifteen states they face tightening legal restrictions on emissions. If your state has not joined that coalition, a good political target is to push your lawmakers into doing so. At the local level, cities and school districts also need to be pushed to stop buying gasoline cars and diesel buses for their fleets, switching to electric vehicles as quickly as possible.

The new urbanist agenda that we described in Chapter 5 requires citizen action, too. Covid-19 taught us something important: people prefer streets that are not wholly devoted to cars. Restaurants expanded into parking spaces, which suddenly served dozens of citizens instead of one SUV driver. Pocket parks allowed kids to burn off energy, and moms and dads to socialize. Bicycle riders felt safer, as lanes were blocked off for them. Farmers markets sprouted, as did tiny libraries, bike parking, and a calmer, more humanistic street life. We have to lock that in and assert that urban public space should serve more than just one need. Citizens also must speak up in favor of imposing congestion charges in cities.

New investments to upgrade American public transit systems are imperative. In 2021, Congress allocated money to rescue transit systems from the financial black hole created by the coronavirus, but further investments to upgrade these systems are still needed. President Biden is pushing that idea, but if the federal government does not come through, states and cities must tackle this problem. One way would be to pass new local taxes, such as a half-penny sales tax to fund a transit construction program. Los Angeles and some other cities have done that, using the tax proceeds to sell bonds that will allow them to move forward with tens of billions of dollars of projects. Cities and their citizens need to think hard, though, about whether shiny new trains or streetcars are the best use of this money. The high-tech bus systems that we described earlier, called Bus Rapid Transit systems, will be a better use of new money in many places. When measures like new transit taxes are under consideration or on the ballot, climate voters need to organize in favor of them.

Perhaps the hardest problem for cities to tackle will be to sharply curtail or end single-family zoning. This practice has done so much to drive urban sprawl in the United States, and the resulting high emissions from automotive transportation. Urban living that limits or eliminates car use is simply not possible in places where it is illegal to build at urban densities. One of the answers is to revise the zoning codes so that "infill" development becomes possible, with single-family houses gradually supplemented with "Granny units," or replaced by duplexes, triplexes, or small apartment buildings. This step is also critical for limiting the runaway home prices that afflict so many of our major cities.

Serious decarbonization of heavy industries is still years away; it is not even clear yet what the most important technological solutions will be. But a few possibilities have begun to appear, which opens another set of political targets for green citizens, as we discussed in Chapter 7. The most critical step right now is sending signals to the marketplace: if a company develops a low-emissions substitute for some high-emissions industrial product, that company needs to know it will find buyers. A prime approach is the Buy Clean ordinances that California and a handful of other states have adopted in recent years. In essence, a state uses its own purchasing power to pull the market toward cleaner products. The goal is to launch these greener industrial products onto the learning curve, so that they get cheaper over time and eventually conquer the market. The federal government has decided to follow California's lead; the more states join in, the faster things will go.

To help repair our relationship with the land, citizens can vote with their wallets. As we discussed in Chapter 6, we urge people who eat meat to cut back on it in their diets, and to give the new meat substitutes a try. If you buy wood or paper products, for yourself or for a company, look for certification by the Forest Stewardship Council, a credible group that is attempting to police the destruction of tropical forests. The federal government needs to launch a more ambitious program to stop illegal wood at the border, but we think state governments need to be pressed to get involved, too. As a pilot program, we would like to see a handful of states start genetic testing for wood products being sold in their states, with a particular focus on stopping the sale of tropical hardwoods.

As this incomplete list illustrates, the political opportunities for citizens to intervene in the climate issue are numerous, and some of them are as close as City Hall. Of course, no single citizen can be expected to advocate for every item on that list; it would be a full-time job. But we believe that every green citizen has the power to contribute—by voting, certainly, but also by writing a letter or making a phone call, by showing up to speak at a hearing, by saying yes instead of no when a renewable-energy project is proposed in their area.

And yet, even if you are willing to do those things, a fair question presents itself. If we know this is going to be a long slog, if we are working toward a distant goal that many of us will not live to see come to pass, how are we to keep our spirits up for the fight?

"To Repair the World"

Not far from the banks of the upper Potomac River, in a charming West Virginia town called Shepherdstown, stands a handsome Presbyterian church. Built in 1836 out of red bricks in a plain, graceful meeting-house style, the church has evolved over the years into one of the most forward-looking in the state. It meets in fellowship with local Muslims, it is friendly with the other churches in town, and it welcomes gay congregants. Its members also think about their footprint on the Earth and how to lighten it.

Among the parishioners is a couple named Nathan and Mary Anne Hitt, who have a beautiful daughter named Hazel. The Hitts are what you might call an environmental power couple. Dr. Hitt, who goes by the nickname Than, is a fisheries biologist for a research arm of the United States government. For a long time, Mary Anne ran national advocacy campaigns for the Sierra Club, and she recently joined a new foundation called Climate Imperative.

When the Hitts moved to town and joined the Presbyterian church, an idea was already percolating in the congregation. The old building is a historic structure, but the church happened to have a recently constructed annex nearby. Would it work to cover that roof in solar panels? The tale of how Than Hitt helped the church get it done is long and winding, but when the work was at last completed in 2014, the Shepherdstown church made news across the region. As an ebullient crowd dedicated the largest community solar project in West Virginia, a state where coal had reigned for a century, the pastor, Randy Tremba, conveyed the sentiments of his congregation.

"The Earth and its wondrous web of life is clearly not of our own making," he declared. "It's a gift and we know it, or should. It's our job to sing its praises, photograph its wonders, and treat it with utmost gratitude and respect. It is sacred. Every step we take is on holy ground."

As news organizations interviewed Dr. Hitt, he went beyond the facts at hand to offer a broader argument. He believes the national discussion about the climate crisis often sounds too clinical—that it is, in effect, missing something. In his view, the cold scientific facts are not necessarily enough to move people to action; indeed, they are so frightening that they can be paralyzing.

"The science is critical, but it's not enough," Mr. Hitt said in one interview. "Science is like a compass. It can tell us where north is, but it can't tell us if we want to go north. That's where our morality comes in."

Despite the reactionary tendencies of many evangelical churches in the United States, Christianity in this country has spawned its own branch of the environmental movement, sometimes operating under the banner "creation care." Thousands of churches have started programs or taken steps toward environmental stewardship, and hundreds have put up solar panels or helped their members learn ways to conserve energy. When Than and Mary Anne Hitt put solar panels on the roof of their own house, a half dozen neighbors in Shepherdstown followed suit. The significance of the project ultimately extended beyond Shepherdstown. The ribbon-cutting was the genesis of a new company, Solar Holler, that went on to become a major Appalachian solar developer, providing dozens of union jobs across West Virginia.

A year after the events in West Virginia, in one of the most profound moments of the climate crisis, Pope Francis issued a powerfully written encyclical directly addressing the threat, the first time the Catholic Church had done so. The pope declared that the Earth "now cries out to us because of the harm we have inflicted on her by our use and abuse of the goods with which God has endowed her."

The idea that each generation of humanity has a duty to care for the interests of future generations is an ancient sentiment, present in many of the world's religions. You do not need to be a believer, though, to feel that way. Sometimes we grow impatient when we hear the tempestuous arguments of economists about whether it is worth paying a few dollars more or less to stave off the worst of the climate crisis. The fractious disputes of engineers arguing for one type of power plant over another can leave us cold. It is not that these arguments are useless, but they do somewhat miss the point. Climate change is not just a pressing physical emergency. It is a *moral* emergency.

The people who originally dug coal, oil, and gas out of the ground did not understand the potential consequences. Even when the use of these fuels escalated many-fold in the middle decades of the twentieth century, even when the first measurements of rising carbon dioxide were made in the 1950s, people did not know. *Now we know.* The scientists among us have issued a clarion warning about the risks we are running. Like Adam and Eve in the Garden of Eden, we have eaten of the forbidden fruit: we are cursed with knowledge.

To continue using these fuels with abandon—to fail to make every effort to wean ourselves from them—is to inflict a problem on future generations that they may have no way to solve. It is entirely possible that children being born today will live to see the worst of it. We may be dooming them to a world riven by conflict as refugees in the hundreds of millions flee rising heat and rising water. We may be condemning the world's great cities to drown. We are endangering much of mankind's cultural heritage. We have no right to wreck the planet for future generations. To us, that means the environmental cleanup we have outlined in this book is not just wise or rational; it is a duty.

One religion in particular offers a framework that we find helpful for thinking about the road ahead. Rooted deep in the Jewish tradition is a sense of obligation captured in the Hebrew phrase תיקון עולם, transliterated as *tikkun olam*, the duty to repair the world. It enjoins the Jewish people to care not just for themselves, but for society at large. Across generations, acts of great charity and beneficence have been undertaken in the pursuit of *tikkun olam*. Not the least of these were the risks incurred by young Jews in the 1960s as they marched in the segregated South to support Black Americans in their fight for civil rights; some of them never made it home. A passage from the collection of ethical teachings known as the Pirkei Avot speaks directly to our hearts as we contemplate the future. "It is not your responsibility to finish the work of perfecting the world," Rabbi Tarfon is quoted as saying, "but you are not free to desist from trying."

Recently, a reporter whose journalism we admire greatly, Sammy Roth of the *Los Angeles Times*, cited that passage as he tried to console Californians battered by climate-induced calamities like wildfires and choking smoke. Mr. Roth wrote: "When I experience climate despair—and I'm guessing this is true for many of you—it stems from the feeling that there's nothing I can do, or that I'm not doing enough. But the reality is I'm only one person. There's only so much I can ask of myself. Just because I'm not all-powerful doesn't mean I'm powerless."

We citizens are only just realizing that, together, we have the power to force change in our politics and our economy. We have always had it, but it has taken us a while to understand the climate problem, to grasp what can and must be done, and to wrap our minds around the urgency of it.

Now, as clarity dawns, we all have work to do.

Acknowledgments

This book distills the ideas of hundreds of people. They have been our guides, muses, and cheerleaders as we tried to think through all the ways that citizens can tackle the emissions that threaten to wreck the Earth. *The Big Fix* is theirs more than ours, though the errors and omissions in this work are ours alone.

Some key backers gave us not only moral but financial support. John Doerr, the Silicon Valley venture capitalist, and his wife, Ann, awarded us an early grant to launch the project. The analysis shop that Hal runs in San Francisco is called Energy Innovation, and it is supported by a network of donors who have chosen to put their money into trying to solve the world's hardest problem. We thank the Doerrs in particular, as well as the other donors who have supported Energy Innovation for years.

In seeking to understand the learning curve and the critical role it has played in the development of modern energy technologies, we benefited from discussions with some of the world's best scholars on the topic. Robert Williams of Princeton was an early and able advocate for studying learning curves in energy technology, and he has been an advisor to Hal for decades. We particularly thank J. Doyne Farmer of the Institute for New Economic Thinking at the University of Oxford, as well as two young scholars in that institute, François Lafond and Rupert Way. Through the generosity of the Oxford Martin School and the gracious supervision of the great economist Cameron Hepburn, Justin was able the spend the Michaelmas term at Oxford in 2018; we thank the philanthropist Lillian Martin for her continued enthusiasm and support of the visitors' program there. Dr. Hepburn, Eric Beinhocker, Myles Allen, Raymond Pierrehumbert, and others at Oxford offered helpful discussions. We thank Julian Garnsey and RWE Renewables for their generosity in arranging a tour of their wind farms off the English coast. At the Rochester Institute of Technology, Eric Hittinger and Eric Williams have done important work on technological learning rates, and they were kind to grant us interviews. Greg Nemet at the University of Wisconsin helped us understand his findings about how solar panels in particular marched so far down the learning curve.

In thinking about the electrical grid, we drew heavily on the work of the team at Energy Innovation that focuses on that problem. Robbie Orvis, Eric Gimon, and Michael O'Boyle are among the smartest people in the country on how we can engineer a grid cleanup. Sonia Aggarwal was the leader of the team and a big influence in the early stages of the book, before she took a job in the White House. We also benefited from the work of two of the smartest scholars seeking to model the grid of the future, Jesse Jenkins at Princeton University and Christopher Clack at Vibrant Clean Energy. We had helpful discussions with Ric O'Connell and Taylor McNair at GridLab. We drew on the work of Amol Phadke, David Wooley, and their colleagues at the Goldman School of Public Policy, University of California, Berkeley, in publishing *2035: The Report*, a visionary document about how to clean up both the electric grid and the transportation system. Lola Spradley, former speaker of the Colorado House of Representatives, was generous with her time in helping us understand the fight over the clean-power requirement passed by the voters of Colorado. Craig Cox kindly shared his recollections and his records in helping us reconstruct those events, and Matt Baker granted several interviews about that era, as well. We thank David Osterberg in Iowa, Nancy Rader in California, and Michael Noble in Minnesota for interviews about how similar requirements came to pass across the country. Alan Nogee, who had a long career with the Union of Concerned Scientists, was helpful in understanding this history, too. We thank Ben Fowke and his crew at Xcel Energy for discussions about their plans regarding clean power, and Colleen Mahoney for opening doors at the utility. The folks at Portland General Electric spent hours of their time helping us understand their efforts regarding demand-response programs; we thank, in particular, Steven Corson, Dave Robertson, Brett Sims, Larry Bekkedahl, Josh Keeling, Andrea Platt, and Rebecca Brisson. Ryan Hledik at Brattle, the consulting firm, helped us understand that group's findings about the broad national potential of demand response. David Pomerantz and his organization, the Energy and Policy Institute, are vital watchdogs of the utility industry, and they filled us in on the details of the corruption scandals in Ohio and Illinois, as well as the broader issue of malign utility influence in state politics. Bryony Worthington, who sits in the House of Lords and is one of the most effective environmental advocates in Britain, helped us understand climate politics there. Jerry Brown, who served California so ably as a visionary governor, shared recollections of his early efforts on alternative energy. Beyond these experts, we benefited and were inspired by the several hundred advocates working on the nitty-gritty particulars of utility regulatory reform. Their work is profoundly important.

For discussions about buildings and how to fix them, we particularly thank Nate Adams of HVAC 2.0, who is seeking, with a partner, to enlist heating and air conditioning contractors across the country in cutting emissions. Mr. Adams has become famous on Twitter as "Nate the House Whisperer." We thank Mr. Adams' clients, John and Sharon Poucher, for allowing us to interview them and reconstruct their travails with their house in Ohio. For discussions about building performance standards, we thank Cliff Majersik of the Institute for

Market Transformation, and for similar discussions about the bizarre politics of model building codes, we thank Bill Fay of the Energy-Efficient Codes Coalition. At the New Buildings Institute, Ralph DiNola, Kim Cheslak, Jim Edelson, Alexi Miller and others gave us the benefit of their thinking about how America can construct buildings fit for the twenty-first century. Jacob Corvidae of the Rocky Mountain Institute shared his thinking on that subject, too. Duane Jonlin kindly filled us in on his work in Seattle, as well as his national advocacy for better buildings. David Kaiserman helped us understand the commercial realities of putting up greener homes. The staff of the Appliance Standards Awareness Project was most helpful, especially Andrew deLaski, the executive director. We thank Alan Meier of the Lawrence Berkeley National Laboratory for helping us understand his work in getting the world's power transformers cleaned up. We thank Holmes Hummel of Clean Energy Works for discussions about the Pay as You Save concept.

Seminal work on the transport system is carried out by the International Council on Clean Transportation. We thank Drew Kodjak and his crew for helpful discussions over many years. John German of I.C.C.T. granted us an interview and we drew on his analysis of the upgrades Toyota made in the 2018 Camry. We thank the dynamic Clemans-Cope sisters, Rosa and Eleanor, for filling us in on their efforts to push the Montgomery County School Board in Maryland to adopt electric buses and other climate measures. Our dear friend Elinor Burkett helped us understand her reporting on Golda Meier's state of mind during the Yom Kippur War. Chris Nelder, an expert on electric mobility and the host of an excellent podcast called *The Energy Transition Show*, was most helpful to us in understanding the problem of charging infrastructure.

Jonas Eliasson is Sweden's best transportation economist and one of the world's leading thinkers about urbanism; he was generous with his time in helping us understand the Stockholm congestion charge and how it came about. We thank Nuzulack Dausen, an enterprising journalist in Dar es Salaam, Tanzania, for the interviews he conducted there on our behalf, and we thank his interviewees: Emmanuel Herman, Ronald Rwakatare, and others. Alix Burgun did similar work for us in Paris, helping us to understand the urban reform program that has slashed automobile traffic in that city. We thank Brent Haddad for discussions about the history of city repair in Vancouver.

Robin Chazdon of the University of Connecticut helped us understand the many issues swirling around tropical forests and land use. Glenn Hurowitz of Mighty Earth and Dan Nepstad of the Earth Innovation Institute granted us interviews for this book and have long served as our guides on saving the forests. The great biologist and conservationist Tom Lovejoy, who was helpful to us over many years in understanding the risks to tropical forests, sadly died in late 2021, as this book was being completed; the world lost a giant.

As we thought through industrial emissions and how to solve them, we drew heavily on the work of Jeffrey Rissman of Energy Innovation, who has become a national leader in analyzing that problem. We also thank Rebecca Dell of ClimateWorks, who granted us interviews to help us understand her excellent, in-depth report on industrial emissions. We thank Patrick Waldron, then of Evraaz Steel, for opening doors and allowing us to tour the company's great steel plant in Pueblo. Governor Jared Polis of Colorado granted us multiple interviews to discuss his vision of a clean-energy economy in that state, including his support for the solar farm that will help to clean up the power emissions associated with the steel plant. We thank Ellen Yui and Tom Schuler for a tour and interviews at Solidia, the New Jersey company that has brought lower-emissions cement to market. Alex Jackson of the Natural Resources Defense Council gave us the benefit of his thinking on industrial emissions.

At the Nuclear Energy Institute in Washington, we thank Matthew Wald and Mary Love for opening doors, and John Kotek and Marc Nichol for their generosity in granting interviews. We thank Chris Greig, now at Princeton, for his help in understanding the abortive ZeroGen project in Queensland. We thank Tim Latimer at Fervo

Energy for discussions about the future of geothermal energy, and we also wish him the very best of luck in bringing his company's first project to commercial operation. We thank Joseph Moore of the University of Utah and his colleagues for a tour of the federally funded geothermal research site in Utah and for helpful discussions there.

Bill Miller of the Anschutz Corporation has a remarkably broad view of the American energy landscape. He was generous with his time in granting us multiple interviews and a tour of the Overland Trail Cattle Company, the Wyoming ranch that is the prospective home of the nation's largest wind farm. We thank Adam Browning, who spent many years at Vote Solar, for helpful discussion and for his pithy motto: "If your plan involves Congress, it's a bad plan." We thank Nathan and Mary Anne Hitt for recounting how solar power penetrated a small West Virginia town, and more importantly, for their broad views about the ethics of tackling climate change. We thank Sammy Roth of the *Los Angeles Times* for his piece pointing us toward the quote from Rabbi Tarfon, and for his superb coverage of climate and energy in that newspaper.

With deep regret, we are omitting the names of more than 300 people from around the world who granted us interviews but wound up not being quoted in the manuscript. To have to leave them out due to space constraints is painful, because every single one of them influenced our thinking. If you are in this large group of our interlocutors, we hope you will find your ideas in the book even if your name is not.

We are both in our early sixties, which means we have had a lifetime of discussion with colleagues and friends about the role of science in society. We simply could not have written this book without that personal history. Over many years, our friends sharpened our wits, turning us into better thinkers. Our professional colleagues and interlocutors did the same, often under the pressure of deadlines and other exigencies.

The *New York Times* in 2010 gave Justin the chance to cover the climate beat, replacing the great reporter Andy Revkin. That assignment allowed him to learn the scientific background and the policy issues swirling around climate change. We thank the *Times* editors who were involved in that coverage and helped to make it better: Glenn Kramon, Erica Goode, Sandy Keenan, Adam Bryant, Celia Dugger, Mary Ann Giordano, Matt Purdy, Bill Keller, Jill Abramson, Dean Baquet and others. Barbara Strauch, the science editor of the *Times* from 2011 to 2015, had a profound influence on the coverage; tragically, she died in 2015 of breast cancer.

Justin thanks several friends who have been interlocutors and guides over the course of a lifetime. Dan Schrag of Harvard University inspired Justin's switch from covering biology to covering climate change, and for many years now has graciously shared his encyclopedic knowledge of climate science and policy. A fellow reporter, Jeff Tollefson, helped to hone many of Justin's earliest ideas for how to cover this subject. David Von Drehle, Peter Slevin, Elinor Burkett, Phil Kuntz, Deborah Sontag, Marianne Gardner, Michael Finney, Gypsy Achong, Paul Crockett, Thomas Priestly, Leonard Laguna, Ross Whittier, Nicholas Johnston, Neil Irwin, Steve Goldenberg, Ellen McCarthy, Fran Brennan, Scott Price, Randall Morris, Chuck Reece, and Tom Lee have also been intellectual guides for decades. Mr. Morris is a working farmer in Georgia; he may not agree with everything we say in our chapter on food and land, but he certainly helped us understand the real difficulties of food production. John O'Leary, David Snydacker, Zach Pierce, and Tyler Norris are newer friends who are directly involved in the energy transition, and whose advice has become indispensable. Jack Price, now on active duty in the United States Marine Corps, has grown up to become one of the most intelligent interlocutors a person could ever have.

Justin also thanks the friends who put up with his couch surfing as he traveled the country, working on this book. David and Benita Nelson of Statham, Georgia, opened their home and hearts, as did Phil Kuntz and Patti Kohaut Parent in Rumson, New Jersey; Phil's sons Liam and McDonagh kindly tolerated an intruder barging into their house for months. No family was kinder, though, than that of Dr. Finney and Dr. Achong in San Francisco; they and their children, Jacqueline and Eli Finney, not only put Justin up in their basement when the project started, they ultimately had to shelter him through a global pandemic.

For formative, early discussions about climate change and how to solve it, Hal thanks Ruth Adams, Ralph Cavanagh, José Goldemberg, Denis Hayes, Frank von Hippel, John Holdren, and Amulya Reddy—along with the many others who have guided and influenced his thinking.

Putting this book together required an entire team. If Tom Shroder is not the best freelance book editor in America, we don't know who might be. Amanda Myers and Mark Silberg did heroic work fact-checking the manuscript, running down leads and setting up interviews. They were both overqualified for the work, as proven by subsequent events: Amanda joined a startup called WeaveGrid, which is pushing transport electrification in new directions, and Mark recently became the top climate advisor to Governor Jared Polis of Colorado. We thank Jules Kortenhorst and Leia Guccione of the Rocky Mountain Institute for lending Mark to us when we needed him most, and we thank Amory Lovins and the entire RMI team for their decades of stellar work on the energy transition.

Our agent, Howard Yoon of the Ross Yoon Agency in Washington, D.C., was both a tough editor and a fount of brilliant ideas; this book would not exist without his efforts, and we thank him deeply.

Energy Innovation is lucky to have a support staff capable of juggling too many moving parts. We particularly thank Christina Fernandes, Clarissa Lopez, and Giselle Stern for their work getting contracts signed, bills paid, travel arranged, and much more; not the least of their feats was shipping an entire research library across the country.

We must thank America's best publishing house, Simon & Schuster, for turning this book from a spark of an idea into a reality. Jonathan

Karp, one of the nation's great editors, is ever alert to the zeitgeist; he sensed that the time might be coming for a book like this and reached out to us in 2017 to solicit a proposal. Priscilla Painton guided this project to completion, and our editor, Megan Hogan, was skillful and wise beyond her years in sharpening the manuscript. We thank the eagle-eyed proofreaders and copyeditors who caught so many errors, large and small: Kathryn Higuchi, Rick Willett, and Anthony Newfield; plus indexer Charles Newman. We thank the book's designer, Kyle Kabel, for the keen sense of style with which he imbued the work.

This book ate a great deal of our personal time. Hal thanks Heather and his children Jeremy, Mariah, and Thea—for ethical inspiration, and in practical terms, for tolerating his long absences, his travel schedule, and his obsession with the climate problem. Justin thanks his brother, Jason Gillis, and Jason's family—Jody; Liz and Hayden; Ashlee and Taylor; and Alexis, Jay, Avery, and Avonlea—for tolerating an uncle whose visits in recent years have been too few and too short. Jason builds houses for a living, to high environmental standards; if readers find any common sense in our buildings chapter, he is likely the source of it.

We are lucky that both our mothers are thriving and will be able to hold this book in their hands. We thank Connie Harvey and Reta Herndon Frisancho for their love, their patience, their intelligence—and for setting each of us onto a meaningful path in life.

Connie has been a singularly effective champion of the wilderness for most of her lifetime, and for that, we both thank her.

Notes

Introduction

1 *Ten thousand firefighters and other workers:* Tahir Husain, *Kuwaiti Oil Fires: Regional Environmental Perspectives,* 1st edition (Oxford: Pergamon, 1995), 67.

1 *the Kuwaiti oil fires consumed only 2 percent:* A wide range of estimates was published about the amount of oil burning in the fires, but the study that we judge to be the most reliable put the figure at 4 million barrels per day, equivalent to a little less than 2 percent of global fossil-fuel consumption today. See Husain, 83.

Chapter 1: The Learning Curve

10 *may ultimately be capable of supplying a large fraction of the world's electrical power:* "Offshore Wind Outlook 2019: World Energy Outlook Special Report" (International Energy Agency, 2019), 98.

10 *they ordered their utilities to spend four or five times the prevailing electricity cost:* John Aldersey-Williams, Ian D. Broadbent, and Peter A. Strachan, "Better Estimates of LCOE from Audited Accounts—A New Methodology with Examples from United Kingdom Offshore Wind and CCGT," *Energy Policy* 128 (May 2019): 25–35, https://doi.org/10.1016/j.enpol.2018.12.044. Historic UK wholesale electricity market prices compiled from Elexon and National Grid by the Institution of Civil Engineers, April 2017.

10 *But lately, contracts have been signed:* Malte Jansen et al., "Offshore Wind Competitiveness in Mature Markets Without Subsidy," *Nature Energy* 5, no. 8 (August 2020): 614–22, https://doi.org/10/1038.s41560-202-0661-2.

11 *Advanced batteries, which power electric cars:* Tifenn Brandily, "1H 2021 LCOE Update," *Bloomberg New Energy Finance*, June 23, 2021.

11 *When highly efficient light bulbs:* "Energy Savings Forecast of Solid-State Lighting in General Illumination Applications" (U.S. Department of Energy Office of Energy Efficiency and Renewable Energy, December 2019), 17, https://www.energy.gov/sites/prod/files/2019/12/f69/2019_ssl-energy-savings-forecast.pdf.

12 *Yet they are but a fraction of the offshore turbines Britain has built:* "UK Wind Energy Database (UKWED)," *RenewableUK*, accessed August 31, 2021, https://www.renewableuk.com/page/UKWEDhome/Wind-Energy-Statistics.htm.

13 *"The amount of interest you get talking to someone about offshore wind":* Julian Garnsey, project director, RWE Renewables, personal interview, July 2019.

14 *The turbine size will jump another 50 percent:* "The Offshore Array," *Triton Knoll*, accessed August 31, 2021, https://www.tritonknoll.co.uk/about-triton-knoll/the-offshore-array/.

14 *Vestas has announced it will build a turbine:* "Vestas Launches the V236-15.0 MW to Set New Industry Benchmark and Take Next Step Towards Leadership in Offshore Wind," *Vestas Company News*, February 10, 2021, https://www.vestas.com/en/media/company news.

15 *It was too close to shore and was thwarted:* Jon Chesto, "R.I.P., Cape Wind," *Boston Globe* , accessed September 10, 2021.

16 *All through the 1920s, the company used Mr. Wright's method:* T. P. Wright, *Articles and Addresses of Theodore P. Wright*, vol. 2 (Buffalo, N.Y.: Cornell Aeronautical Laboratory, Inc., 1961), 32.

17 *"Our potential enemies had selected the weapon":* T. P. Wright, *Articles and Addresses of Theodore P. Wright*, vol. 3 (Buffalo, N.Y.: Cornell Aeronautical Laboratory, Inc., 1961), 50.

17 *The paper he finally published:* T. P. Wright, "Factors Affecting the Cost of Airplanes," *Journal of the Aeronautical Sciences* 3, no. 4 (February 1936): 122–28, https://doi.org/10.2514/8.155.

17 *By the peak of the Model T in the mid-1920s:* Chris Brancaccio, "Encyclopedia," Model T Ford Club of America, accessed September 16, 2021, https://www.mtfca.com/encyclo/.

18 *Its most famous innovation was the moving assembly line:* The Reminiscences of Mr. W. C. Klann, September 1955, Benson Ford Research Center, https://

cdm15889.contentdm.oclc.org/digital/collection/p15889coll2/id/7804. On page 51 of his reminiscences, Mr. Klann, an engineer at Ford, describes the trip to Chicago that inspired the assembly line. Henry Ford makes the same point in his 1922 autobiography: "the idea came in a general way from the overhead trolley that the Chicago packers use in dressing beef." See Henry Ford, *My Life and Work: An Autobiography of Henry Ford* (United States: Greenbook Publications, 2010), 55.

19 *Mr. Wright finally found his answer:* T. P. Wright, *Articles and Addresses of Theodore P. Wright*, vol. 2 (Buffalo, N.Y.: Cornell Aeronautical Laboratory, Inc., 1961), 34.

20 *By late 1944, the labor requirement had been slashed:* "B-17 Production and Construction Analysis" (Air Materiel Command, May 29, 1946).

21 *To this day, scholars still study:* See, for instance: François Lafond, Diana Seave Greenwald, and J. Doyne Farmer, "Can Stimulating Demand Drive Costs Down? World War II as a Natural Experiment," *SSRN Electronic Journal* 2020, https://doi.org/10.2139/ssrn.3519913.

21 *"We know that Franklin Roosevelt was not buying more airplanes":* J. Doyne Farmer, director of complexity economics at the Institute for New Economic Thinking at the Oxford Martin School, University of Oxford, personal interview, November 2018.

21 *Researchers have established that Moore's Law and Wright's Law are closely related:* See the comparison between Wright's Law and Moore's Law in Béla Nagy et al., "Statistical Basis for Predicting Technological Progress," *PLOS ONE* 8, no. 2 (February 28, 2013): e52669, https://doi.org/10.1371/journal.pone.0052669.

22 *But proponents of the idea went overboard:* Bruce D. Henderson, *Henderson on Corporate Strategy* (Cambridge, Mass: Abt Books, 1979), 12, 14, 18; Carl W. Stern, George Stalk, and Boston Consulting Group, eds., *Perspectives on Strategy: From the Boston Consulting Group* (New York: J. Wiley, 1998), 18, 22. Henderson and his colleagues at the Boston Consulting Group attempted to apply experience curves to multiple domains of business strategy.

22 *Worse, running a manufacturing business with a healthy learning rate:* This is essentially what happened to the Ford Motor Company. When General Motors, under the leadership of Alfred P. Sloan, put superior cars on the market in the mid-1920s, sales of the Model T started to fall. Henry Ford no longer enjoyed the rapid cost declines that had been the wind in his sails,

and by the end, was losing money on the car. The Model T had been his grand success, but he loved it too much and clung to it too long. Ford Motor lost its leadership of the automotive industry to General Motors, never to regain it. See: William J. Abernathy and Kenneth Wayne, "Limits of the Learning Curve," *Harvard Business Review* (September 1, 1974), https://hbr.org/1974/09/limits-of-the-learning-curve.

24 *After a short stint in jail, he fled California:* Peter Asmus, *Reaping the Wind: How Mechanical Wizards, Visionaries, and Profiteers Helped Shape Our Energy Future* (Washington, D.C: Island Press, 2001), 111–114; Robert W. Righter, *Wind Energy in America: A History* (Norman, Okla: University of Oklahoma Press, 1996), 87–90; Steven Lech, "Back in the Day: Wind Machine Predated Iconic Desert Turbines," *Press-Enterprise* (Riverside, Calif.), April 12, 2015; Nicole C. Brambila, "Harnessing Local Wind for Energy Not a New Idea," *Desert Sun*, May 17, 2009; David S. Smith, "Pass Area Seen Ideal for Wind Energy Study," *Desert Sun*, June 16, 1976, https://www.newspapers.com/image/747668634/; Ralph Hinman, "Saga of the 'Wind Machine Man,'" *Press-Telegram* (Long Beach, Calif.), January 13, 1974, https://www.newspapers.com/image/706705859/.

24 *Actually, the notion of generating electricity from the wind:* Although wind chargers took off as a commercial product in the 1920s, the idea had roots deep in the nineteenth century, and many prototypes were built then. See the early chapters of Robert W. Righter, *Wind Energy in America: A History* (Norman, Okla: University of Oklahoma Press, 1996).

24 *This new technology had soared in popularity:* Christopher H. Sterling and John M. Kittross, *Stay Tuned: A History of American Broadcasting*, 3rd ed., LEA's Communication Series (Mahwah, NJ: Lawrence Erlbaum Associates, 2002).

24 *Startup companies plied the countryside:* Ronald R. Kline, *Consumers in the Country: Technology and Social Change in Rural America, Revisiting Rural America* (Baltimore, Md.: Johns Hopkins University Press, 2000).

24 *one of Franklin D. Roosevelt's New Deal programs delivered nearly universal access:* Contrary to popular impression, access to electricity is still not an entirely solved problem in America. Fifteen thousand families in the Navajo Nation, a third of households there, still do not have power. Volunteer efforts are underway to get these homes connected.

25 *"And hence it proved that at some future time":* Palmer Cosslett Putnam, *Power from the Wind* (New York: Van Nostrand Reinhold, 1974), xi.

26 *Aside from pollution and other near-term problems:* Wilson Clark, *Energy for Survival: The Alternative to Extinction*, 1st edition (Garden City, NY: Anchor Press, 1974).

26 *Looking askance at some of the ideas emanating from the state capitol:* Royko later regretted the nickname and pleaded with his audience to stop using it. See Mike Royko, "Time to Eclipse 'Moonbeam' Label," *Chicago Tribune*, September 4, 1991.

27 *The government subsidies and guarantee of a market:* For a history of this era, see chapter 10 of Robert W. Righter, *Wind Energy in America: A History* (Norman, Okla: University of Oklahoma Press, 1996).

27 *The French locked in on nuclear power as the solution:* Arnulf Grubler, "The Costs of the French Nuclear Scale-Up: A Case of Negative Learning by Doing," *Energy Policy* 38, no. 9 (September 2010): 5174–88, https://doi.org/10.1016/j.enpol.2010.05.003.

28 *That design had already been shown to work:* A rotor with three blades was the standard design of the best wind chargers built in America, those of the Jacobs Wind Electric Company. Marcellus and Joe Jacobs tried two blades but found that three minimized vibrations. Robert W. Righter, *Wind Energy in America: A History* (Norman, Okla: University of Oklahoma Press, 1996), 90–99.

28 *The project received extensive news coverage:* Many years later, for unrelated reasons, the leaders of Tvind would be prosecuted by the Danish state for tax evasion and other crimes, and some of them were still fugitives from justice at the time this book went to press.

28 *For a while, Denmark was exporting more turbines:* Paul Gipe, *Wind Energy Comes of Age*, Wiley Series in Sustainable Design (New York: Wiley, 1995), 58.

28 *In 1905, Albert Einstein explained what was happening:* "The Nobel Prize in Physics 1921," NobelPrize.org, accessed August 31, 2021, https://www.nobelprize.org/prizes/physics/1921/summary/.

29 *"The scheme of combustion in order to get power":* Elbert Hubbard and Felix Shay, "The Open Road Afoot with the Fra—Thomas A. Edison," in *The Fra: A Journal of Affirmation* 5, no. 1 (East Aurora, N.Y.: Roycrofters, 1910): 1–8, https://digital.library.villanova.edu/Item/vudl:87792.

29 *The magazine wrote that "in the not distant future":* "Magic Plates Tap Sun for Power," *Popular Science Monthly*, June 1931, 41.

29 *An inventor at Bell Labs named Daryl M. Chapin was working:* Gregory F. Nemet, *How Solar Energy Became Cheap: A Model for Low-Carbon Innovation* (London: Routledge/Taylor & Francis Group, 2019), 59.

29 *A writer for the* New York Times *declared:* "Vast Power of the Sun Is Tapped by Battery Using Sand Ingredient," *New York Times*, April 26, 1954.

30 *Seeing potential in this growing industry:* Gregory F. Nemet, *How Solar Energy Became Cheap: A Model for Low-Carbon Innovation* (London: Routledge/ Taylor & Francis Group, 2019), chaps. 5, 9, 10.

30 *In 1988, a NASA scientist, James E. Hansen, warned Congress:* Philip Shabecoff, "Global Warming Has Begun, Expert Tells Senate," *New York Times*, June 24, 1988, sec. U.S., https://www.nytimes.com/1988/06/24/us/global-warming-has-begun-expert-tells-senate.html.

31 *Not until 1993 did somebody finally make the next leap of logic:* Robert H. Williams and Greg Terzian, "A Benefit/Cost Analysis for Accelerated Development of Photovoltaic Technology," *PU/CEES Report No. 281*, Center for Energy and Environmental Studies, Princeton University, October 1993.

32 *Mr. Cameron cheered as Messrs. Yorke and Greenwood:* "The Big Ask: How You Helped Make Climate Change History," *Friends of the Earth*, accessed September 16, 2021, https://friendsoftheearth.uk/climate/big-ask-how-you-helped-make-climate-change-history.

32 *Asked in an interview why he was doing it:* "The Big Ask | KOKO London," accessed September 30, 2021, http://koko.uk.com/listings/big-ask-01-05-2006.

32 *He had won the post by promising:* Guardian Staff, "Full Text: David Cameron's Speech to the Conservative Conference 2005," *Guardian*, October 4, 2005.

33 *Since clean electricity would be needed:* Committee on Climate Change, *Building a Low-Carbon Economy: The UK's Contribution to Tackling Climate Change: The Report of the Committee on Climate Change*, December 2008 (London: TSO, 2008), 173.

33 *British politicians decided that offshore wind:* To be clear, the politicians took a winding path before they finally arrived at their support for offshore wind. See Michael Grubb and David Newbery, "UK Electricity Market Reform and the Energy Transition: Emerging Lessons," *Energy Journal* 39, no. 1 (September 1, 2018), https://doi.org/10.5547/01956574.39.6.mgru.

35 *But at a global scale, power generation represents only 38 percent of carbon dioxide:* In this book, when we give emissions breakdowns, we are generally supplying figures only for carbon dioxide, the most important greenhouse gas and one with a very long lifetime in the atmosphere. Other data sources often combine carbon dioxide with additional greenhouse gases and give "CO_2-equivalent" figures, so the data you see in this book may differ

somewhat from these other publications. The second-most important greenhouse gas, methane, has a much shorter atmospheric lifetime than carbon dioxide. We deal with methane where it is most relevant: in Chapter 6, relating to land use and food production.

Chapter 2: Power Switch

38 *"I'm going: a windmill, making energy instead of pumping water?":* Lola Spradley, personal interview, June 2, 2019.

38 *"I only won by four hundred votes":* David Osterberg, personal interview, September 18, 2021.

40 *The laws generally went through as part of the larger negotiation:* For a detailed explanation of the economic rationale, see Nancy A. Rader and Richard B. Norgaard, "Efficiency and Sustainability in Restructured Electricity Markets: The Renewables Portfolio Standard," *Electricity Journal* 9, no. 6 (July 1996): 37–49, https://doi.org/10.1016/S1040-6190(96)80262-4.

40 *The power companies argued that rates would go up:* "Colorado Renewable Energy Requirement," *Pub. L. No. Initiative 37* (2004), http://www.leg.state.co.us/lcs/ballothistory.nsf/835d2ada8de735e787256ffe0074333d/c29f58efb1bdce268725702000731ec7?OpenDocument.

40 *They passed the measure, Amendment 37:* "Official Publication of the Abstract of Votes Cast for the 2003 Coordinated, 2004 Primary, 2004 General" (Colorado Secretary of State, 2004), 140, https://www.sos.state.co.us/pubs/elections/Results/Abstract/pdf/2000-2099/2004AbstractBook.pdf.

41 *Eventually, thirty of the fifty American states:* Galen Barbose, "U.S. Renewables Portfolio Standards, 2021 Status Update: Early Release" (Lawrence Berkeley National Laboratory, February 2021), https://eta-publications.lbl.gov/sites/default/files/rps_status_update-2021_early_release.pdf.

41 *And the learning curve worked its magic:* Martin Junginger et al., "Onshore Wind Energy," in *Technological Learning in the Transition to a Low-Carbon Energy System* (Amsterdam: Elsevier, 2020), 87–102, https://doi.org/10.1016/B978-0-12-818762-3.00006-6.

42 *And the economics are changing rapidly:* Eric Gimon et al., "The Coal Cost Crossover: Economic Viability of Existing Coal Compared to New Local Wind and Solar Resources" (*Energy Innovation and Vibrant Clean Energy*, March 2019), https://energyinnovation.org/wp-content/uploads/2019/04/Coal-Cost-Crossover_Energy-Innovation_VCE_FINAL2.pdf.

43 *Wind-power companies pay taxes to support local schools:* An analysis by Rocky Mountain Institute found that in 2030 rural communities may receive annual revenues from wind and solar projects exceeding $60 billion, comparable to expected revenues from the top three agricultural commodities (corn, soy, and beef). Even today, annual revenues from solar and wind are nearly the same as soy crop revenue. See Katie Siegner, Kevin Brehm, and Mark Dyson. "Seeds of Opportunity: How Rural America Is Reaping Economic Development Benefits from the Growth of Renewables." Rocky Mountain Institute, 2021. http://www.rmi.org/insight/seeds-of-opportunity.

43 *"We export lots of things, and in our future":* Justin Gillis and Nadja Popovich, "In Trump Country, Renewable Energy Is Thriving," *New York Times*, June 7, 2017, sec. Climate, https://www.nytimes.com/2017/06/06/climate/renewable-energy-push-is-strongest-in-the-reddest-states.html.

43 *If you look only at the past decade, when most of the world's solar panels were installed:* Atse Louwen and Wilfried van Sark, "Photovoltaic Solar Energy," in *Technological Learning in the Transition to a Low-Carbon Energy System* (Elsevier, 2020), 65–86, https://doi.org/10.1016/B978-0-12-818762-3.00005-4.

46 *"That's mainly because the subject is excruciatingly boring":* The author of this passage, David Roberts, is perhaps the clearest writer in America regarding what we must do to dig ourselves out of the climate problem. He has begun publishing his own newsletter at www.volts.wtf, and we commend it to our readers. The quotation is from "Utilities for Dummies: How They Work and Why That Needs to Change." *Grist*, May 21, 2013. https://grist.org/climate-energy/utilities-for-dummies-how-they-work-and-why-that-needs-to-change/.

47 *Getting electricity to the 770 million people:* The estimate of the number of people deprived of electricity is from the International Energy Agency, *World Energy Outlook 2020*, 40.

48 *Even discounting the climate problem:* Karn Vohra et al., "Global Mortality from Outdoor Fine Particle Pollution Generated by Fossil Fuel Combustion: Results from GEOS-Chem," *Environmental Research* 195 (April 2021): 110754, https://doi.org/10.1016/j.envres.2021.110754.

51 *As demand for power ramps up, the grid usually meets it:* Paul Denholm, Yinong Sun, and Trieu Mai., "An Introduction to Grid Services: Concepts, Technical Requirements, and Provision from Wind" (National Renewable

Energy Laboratory, 2019), https://www.nrel.gov/docs/fy19osti/72578.pdf.

54 Entir*e states have gotten rid of most or all of their coal plants:* Figures about plant closures are based on commercial databases compiled by the Sierra Club, shared in correspondence with the authors, September 27, 2021.

54 *That cancels out some of the benefit of switching from coal to gas:* Researchers at Carnegie Mellon University found that a natural gas power plant offers lifetime climate benefits compared to a coal plant over 100 years, even if methane leakage reaches as much as 5 percent. Over twenty years, natural gas plants with 4 percent leakage have a similar climate impact as coal plants, but "are no worse than coal." It is unlikely that the total natural gas system leakage exceeds 4 percent in the United States. See DeVynne Farquharson, Paulina Jaramillo, Greg Schivley, Kelly Klima, Derrick Carlson, and Constantine Samaras, "Beyond Global Warming Potential: A Comparative Application of Climate Impact Metrics for the Life Cycle Assessment of Coal and Natural Gas Based Electricity: Beyond Global Warming Potential." *Journal of Industrial Ecology* 21, no. 4 (August 2017): 857–73.

54 *More than a dozen had closed by the end of 2021:* Form EIA-860 Data, U.S. Energy Information Administration, 2020, https://www.eia.gov/electricity/data/eia860/.

58 *A research outfit called the Brattle Group found:* Ryan Hledik et al., "The National Potential for Load Flexibility: Value and Market Potential Through 2030" (The Brattle Group, June 2019).

59 *Finally, in a landmark study in 2012:* Trieu Mai et al., "Renewable Electricity Futures Study," (Golden, CO: National Renewable Energy Laboratory, 2012.)

59 *More recent studies argue that as much as 90 percent of grid power:* In 2020, a group of researchers from the Goldman School of Public Policy at the University of California, Berkeley, and collaborators found the United States can achieve 90 percent carbon-free electricity nationwide by 2035 at no extra cost to consumers, and without new fossil fuel plants. See: Amol Phadke, Umed Paliwal, Nikit Abhyandkar, Taylor McNair, Ben Paulos, David Wooley, and Ric O'Connell. "The 2035 Report: Plummeting Solar, Wind, and Battery Costs Can Accelerate Our Clean Energy Future." Goldman School of Public Policy, GridLab, PaulosAnalysis, June 2020, http://www.2035report.com/.

60 *"My kids belong to the generation that will be dealing with many of the impacts of global warming":* The citizen testimony from this hearing can be found in

Harriet S. Weisenthal, "Proceeding No. 16A-0396E Public Comment Hearing," Public Utilities Commission of the State of Colorado (Denver, 2018).

61 *Across its territory, Xcel is already getting nearly half its power from clean sources:* "2020 Sustainability Report" (Xcel Energy, June 7, 2021), https://s25.q4cdn.com/680186029/files/doc_downloads/irw/Sustainability/2020-Sustainability-Report-Full.pdf.

61 *Of all the large electric utilities in the United States, an analysis:* Preliminary data from the Rocky Mountain Institute's Utility Transition Hub compared utility clean-energy commitments to an emissions pathway limiting global warming to 1.5 degrees Celsius, the stretch goal of the Paris Agreement. The analysis found that between 2020 and 2030, Xcel Energy's emissions would be 3.2 percent below the requisite pathway, assuming the company meets its stated goals. Data tracking utility commitments across the United States are available at utilitytransitionhub.rmi.org.

65 *"At first we sent a letter to City Hall":* Claire Thompson, "Meet the Woman Who Shut own Chicago's Dirty Coal Plants," *Grist*, April 15, 2013, https://grist.org/climate-energy/interview-wkimberly-wasserman-nieto-goldman-prize-winner/.

65 *"Forty-one young people laid on the floor":* Ibid.

Chapter 3: Where We Live and Work

67 *"THE BIG ONE," a huge banner headline screamed:* "Hurricane Andrew | Flashback Miami," *Miami Herald* Photos & Archives, August 23, 2016, https://flashbackmiami.com/2016/08/23/hurricane-andrew/.

67 *Two Herald reporters staying in a Florida City hotel:* Don Van Natta, Jr, "Comfort Inn Hero: Fast-Thinking Clerk," *Miami Herald*, August 25, 1992, 4A, https://www.newspapers.com/image/637540434.

68 *The storm destroyed or badly damaged:* Ed Rappaport, "Hurricane Andrew Preliminary Report," National Hurricane Center, December 10, 1993, https://www.nhc.noaa.gov/1992andrew.html; Stanley K. Smith and Christopher McCarty, "Demographic Effects of Natural Disasters: A Case Study of Hurricane Andrew," *Demography* 33, no. 2 (May 1996): 265–75.

68 *To cut corners, some builders had failed:* For documentation of these claims, see the special newspaper section the *Miami Herald* published after Hurricane Andrew: "What Went Wrong," *Miami Herald*, December 20, 1992, sec. Special Report.

68 *Others began pulling out of Florida:* Lynne McChristian, "Hurricane Andrew and Insurance: The Enduring Impact of an Historic Storm" (Tampa, Fla: Insurance Information Institute, August 2012), https://www.iii.org/sites/default/files/paper_HurricaneAndrew_final.pdf.

68 *Florida adopted new laws requiring much tougher building codes:* "2010 Florida Test Protocols for High Velocity Hurricane Zones" (International Code Council, October 2011), Preface, https://codes.iccsafe.org/content/FLTEST2010/preface.

69 *Buildings are one of the nation's largest sources:* "Inventory of U.S. Greenhouse Gas Emissions and Sinks: 1990–2019" (United States Environmental Protection Agency, 2021), ES-13.

70 *"Grassroots" groups have popped up to try to stop the bans:* Miranda Green and Sammy Roth, "They Fought for Clean Air. They Didn't Know They Were Part of a Gas Industry Campaign," *Los Angeles Times*, August 16, 2021, https://www.latimes.com/business/story/2021-08-16/clean-air-gas-trucks-la-long-beach-ports.

77 *"Heat pumps are bisexual air conditioners":* Nate Adams, personal interview, April 2021.

78 *But even at those elevated prices, they are often cost-effective:* Claire McKenna, Amar Shah, and Leah Louis-Prescott, "The New Economics of Electrifying Buildings" (RMI, 2020), https://rmi.org/download/26837/.

79 *The fridge standards save a typical homeowner:* Appliance Standards Awareness Project, in correspondence with the authors, October 2021.

79 *"President Bush was a tech geek":* Dr. Alan Meier, personal interview, August 31, 2021.

80 *Two Japanese scientists and one Japanese American colleague:* "The Nobel Prize in Physics 2014," NobelPrize.org, accessed September 29, 2021, https://www.nobelprize.org/prizes/physics/2014/press-release/.

80 *The market started to grow rapidly:* Brian F. Gerke, "Light-Emitting Diode Lighting Products," in *Technological Learning in the Transition to a Low-Carbon Energy System* (Elsevier, 2020), 233–56, https://doi.org/10.1016/B978-0-12-818762-3.00013-3.

88 *The nation's Realtors may well fight such requirements:* See Erica Myers, Steven Puller, and Jeremy West, "Effects of Mandatory Energy Efficiency Disclosure in Housing Markets" (Cambridge, MA: National Bureau of Economic Research, November 2019), https://doi.org/10.3386/w26436.

Chapter 4: Over a Barrel

93 *"Military exercises now going on in Egypt are larger and more realistic":* "The President's Daily Brief" (Central Intelligence Agency, October 5, 1973), 3.

93 *As Mr. Nixon's Presidential Daily Brief was being completed on that day":* "The President's Daily Brief" (Central Intelligence Agency, October 6, 1973), 3.

94 *According to reporting decades later:* Elinor Burkett, *Golda*, 1st edition (New York: Harper, 2008), 324.

95 *Before October of 1973, only a handful of experts:* Linda W. Qaimmaqami, Adam M. Howard, and Edward C. Keefer, eds., "President's Meeting with His Foreign Intelligence Advisory Board" (United States Government Printing Office, June 5, 1970), 80, National Archives, Nixon Presidential Materials, NSC Files, Box 276, Agency Files, President's Foreign Intelligence Advisory Board, Vol. IV, https://2001-2009.state.gov/documents/organization/113361.pdf.

95 *Even before the Yom Kippur War started, spot shortages:* Meg Jacobs, *Panic at the Pump: The Energy Crisis and the Transformation of American Politics in the 1970s*, 1st edition (New York: Hill and Wang, a division of Farrar, Straus and Giroux, 2016), 37.

95 *This was not some imagined threat:* Joseph Mann, "A Reassessment of the 1967 Arab Oil Embargo," Israel Affairs 19, no. 4 (October 2013): 693–703, https://doi.org/10.1080/13537121.2013.829611.

95 *Gasoline shortages worsened quickly as a frightened public:* Jacobs, *Panic at the Pump*, 59, 65, 79.

95 *Mile-long lines snaked from gas stations:* Ibid., 95.

95 *People spent hours waiting to fill up:* Louise Cook, "Gas Line Violence: Weekends Are Worst," *The Central New Jersey Home News*, February 21, 1974.

95 *Fistfights broke out:* Louise Cook, "Gas Shortage Stirs Violence at Stations," *Casa Grande Dispatch*, February 20, 1974.

95 *Thieves siphoned gasoline out of other people's tanks:* Alan Quale, "Fuel Shortage Brings Changes: Long Lines, Frayed Tempers and Fights," *The Times* (San Mateo, California), February 16, 1974, 15-R.

95 *Angry truckers drove onto the interstates:* Jacobs, Panic at the Pump, 93.

97 *Now we understand, though, that the real risk:* Researchers have established that to limit warming to 1.5 degrees Celsius would require prohibiting the extraction of 58 percent of known oil reserves, 59 percent of natural gas reserves, and 89 percent of coal reserves. Dan Welsby et al., "Unextractable

Fossil Fuels in a 1.5°C World," Nature 597, no. 7875 (September 9, 2021): 230–34, https://doi.org/10.1038/s41586-021-03821-8.

97 *That means as many as 2 billion additional fuel-burning cars:* "Vision 2050: A Strategy to Decarbonize the Global Transport Sector by Mid-Century" (International Council on Clean Transportation, 2020), 11.

97 *As little as 20 percent of the energy is captured:* R. P. Siegel, "The Infinitely Expandable Resource," *Mechanical Engineering Magazine*, August 2020, 48.

100 *Instead, they used the efficiency gains to make cars heavier:* "The 2020 EPA Automotive Trends Report" (United States Environmental Protection Agency, January 2021), 29, https://nepis.epa.gov/Exe/ZyPDF.cgi?Dockey=P1010U68.pdf.

101 *The result was an improvement in gas mileage:* John German, "How Things Work: OMEGA Modeling Case Study Based on the 2018 Toyota Camry" (International Council on Clean Transportation, February 27, 2018), https://theicct.org/sites/default/files/publications/Camry_OMEGA_Working Paper_20180227.pdf.

105 *Cars of all kinds were still toys for rich folks in those days:* Kevin A. Wilson, "Worth the Watt: A Brief History of the Electric Car, 1830 to Present." *Car and Driver*, March 15, 2018, https://www.caranddriver.com/features/g15378765/worth-the-watt-a-brief-history-of-the-electric-car-1830-to-present.

106 *In Norway, a small environmental group:* Jon Henley and Elisabeth Ulven. "Norway and the *a*-h*a* Moment That Made Electric Cars the Answer," *Guardian*, April 19, 2020, sec. Environment, https://www.theguardian.com/environment/2020/apr/19/norway-and-the-a-ha-moment-that-made-electric-cars-the-answer.

107 *"Almost any new technology initially has high unit cost":* The quote is from a blog post by Elon Musk, The Secret Tesla Motors Master Plan (just between you and me), posted August 2, 2006, at tesla.com

108 *Mr. Musk's company has had growing pains:* For instance, he was charged with securities fraud after tweeting to his 22 million Twitter followers that he could take Tesla private for $420 per share, an implicit reference to the slang term for marijuana. That little joke cost the company and Mr. Musk a total of $40 million in fines. See "Elon Musk Charged With Securities Fraud for Misleading Tweets," U.S. Securities and Exchange Commission, September 27, 2018, https://www.sec.gov/news/press-release/2018-219.

108 *In the overall global car market:* "Global EV Outlook 2021" (Paris: International Energy Agency, 2021), https://www.iea.org/reports/global-ev-outlook-2021.

109 *The city has done all this in less than a decade:* Mengnan Li et al., "How Shenzhen, China, Pioneered the Widespread Adoption of Electric Vehicles in a Major City: Implications for Global Implementation," *WIREs Energy and Environment* 9, no. 4 (July 2020), https://doi.org/10.1002/wene.373.

110 *Chris Nelder, an electric car expert:* Chris Nelder, personal interview, January 26, 2022.

110 *Considering the total running costs over the lifetime of a car:* This is a national average; you can get a more exact number for your state by visiting the Department of Energy's "eGallon" Calculator online at https://www.energy.gov/maps/egallon.

111 *"I like to stand up for my community":* Priscilla Totiyapungprasert, "These High School Runners Train in 'Nasty Air,' so They're Working to Clean It Up," *Arizona Republic*, July 29, 2019, https://www.azcentral.com/story/news/local/arizona-environment/2019/07/29/why-these-south-mountain-students-fighting-electric-buses-pollution-clean-air/3104934002/.

111 *The truth is that we need to act as if our house is on fire:* Rosa Clemans-Cope, personal interview, November 4, 2021.

112 *The financial package was attractive enough that Montgomery County ordered:* Steve Mufson and Sarah Kaplan. "A Lesson in Electric School Buses." *Washington Post*, February 24, 2021. https://www.washingtonpost.com/climate-solutions/2021/02/24/climate-solutions-electric-schoolbuses/.

113 *California is also leading a coalition of states:* As this book went to press, the signatories of the heavy-vehicle compact included the governors of California, Colorado, Connecticut, Hawaii, Maine, Maryland, Massachusetts, New Jersey, New York, North Carolina, Oregon, Pennsylvania, Rhode Island, Vermont, Washington, and the District of Columbia. See: "Multi-State Medium- and Heavy-Duty Zero Emission Vehicle Memorandum of Understanding" (2020), https://www.energy.ca.gov/sites/default/files/2020-08/Multistate-Truck-ZEV-Governors-MOU-20200714_ADA.pdf.

116 *"I see the power of solidarity all the time":* Rosa Clemans-Cope, personal interview, November 4, 2021.

Chapter 5: An Urban Planet

118 *He had even heard from his aunt, Gunnel Hahne:* Jonas Eliasson, personal interview, December 3, 2018.

118 *Picking up the newspaper, "I gaped":* Jonas Eliasson, "Lessons from the Stockholm Congestion Charging Trial," *Transport Policy* 15, no. 6 (November 2008): 11, https://doi.org/10.1016/j.tranpol.2008.12.004.

119 *Researchers later confirmed a sharp drop:* Emilia Simeonova et al., "Congestion Pricing, Air Pollution and Children's Health" (Cambridge, MA: National Bureau of Economic Research, March 2018), https://doi.org/10.3386/w24410.

120 *Contrast that with Chicago:* Figures for per capita vehicle miles traveled are from "State & Urbanized Area Statistics—Our Nation's Highways" Federal Highway Administration, United States Department of Transportation, (March 29, 2018), https://www.fhwa.dot.gov/ohim/onh00/onh2p11.htm.

123 *Yes, the construction of 42,500 miles of American interstate highways:* Kenneth T. Jackson, *Crabgrass Frontier: The Suburbanization of the United States* (New York: Oxford University Press, 1985), 249.

124 *According to Johnson, "some city officials expressed the view":* Richard Rothstein, *The Color of Law: A Forgotten History of How Our Government Segregated America*, 1st edition (New York: Liveright Publishing Corporation, a division of W. W. Norton & Company, 2017), 128.

124 *Clayton Nall, a researcher at the University of California, Santa Barbara:* Clayton Nall, *The Road to Inequality: How the Federal Highway Program Polarized America and Undermined Cities* (Cambridge, UK: Cambridge University Press, 2018).

125 *To this day, you can go visit a huge hole in Pasadena:* The authors thank Thomas Priestly, an urban planner and thinker of renown, for giving Justin a tour of this very hole, which is not far from his home in Pasadena.

125 *Sidewalks were being narrowed to make more lanes:* Michele Richmond, "The Etymology of Parking," *Arnoldia* (2015).

126 *Their biking cultures, and the extensive facilities that support them:* For a detailed exposition of how Copenhagen became so bicycle-friendly, see Peter S. Goodman, "The City That Cycles With the Young, the Old, the Busy and the Dead," *New York Times*, November 9, 2019, sec. World, https://www.nytimes.com/2019/11/09/world/europe/biking-copenhagen.html.

126 *And we do mean extensive:* Bicycle Parking Stationsplein Utrecht: Largest in the World," City of Utrecht, accessed September 26, 2021, https://www.utrecht.nl/city-of-utrecht/mobility/cycling/bicycle-parking/bicycle-parking-stationsplein/bicycle-parking-stationsplein-utrecht-largest-in-the-world/.

126 *But this bus station, and the larger system of which it is a part:* The details about the Dar es Salaam B.R.T. system were gathered, and interviews with Emmanuel Herman, Charles Mapuli, and Ronald Kwakatare were conducted, by reporter Nuzulack Dausen in Dar es Salaam in early 2020 while on special assignment for the authors of this book.

129 *Janette Sadik-Khan, the transportation commissioner:* For a detailed account of the Times Square episode and other adventures in urban planning in New York City, see: Janette Sadik-Khan and Seth Solomonow, *Streetfight: Handbook for an Urban Revolution* (New York: Viking, 2016), especially chapter 6.

135 *As Bogotá mayor Enrique Peñalosa says:* "Why Buses Represent Democracy in Action" (TEDCity2.0, Manhattan, NY, September 20, 2013), https://www.ted.com/talks/enrique_penalosa_why_buses_represent_democracy_in_action.

137 *"Hordes of people showed up," reporter Lisa Gray wrote:* Lisa Gray, "Building a Better Block," *Houston Chronicle*, June 28, 2010, sec. Culture, https://www.chron.com/culture/main/article/Gray-Building-a-better-block-1711370.php.

137 *Thanks to that* Houston Chronicle *story:* More details of the Houston story, and other tales of tactical urbanism, can be found in Mike Lydon, Anthony Garcia, and Andres Duany, *Tactical Urbanism: Short-Term Action for Long-Term Change* (Washington, DC: Island Press, 2015).

137 *A tactical urbanist group in Portland, Oregon:* Ibid., 92–101.

139 *The latest version has fifteen freeways on the hit list:* Ben Crowther, "Freeways Without Futures" (Congress for the New Urbanism, 2021), https://www.cnu.org/sites/default/files/FreewaysWithoutFutures_2021.pdf.

Chapter 6: People, Land, and Food

142 *"When the European colonists came to America":* Quoted in: Justin Gillis, "With Deaths of Forests, a Loss of Key Climate Protectors," *New York Times*, October 1, 2011, sec. Science, https://www.nytimes.com/2011/10/01/science/earth/01forest.html.

144 *The human population today is approaching 8 billion:* United Nations, Department of Economic and Social Affairs, Population Division, *World Population Prospects Highlights, 2019 Revision Highlights, 2019 Revision*, fig. 1, Population size and annual growth rate for the world: estimates, 1950–2020, and medium-variant projection with 95 per cent prediction intervals, 2020–2100."

145 *The scholar Vaclav Smil calls this human addiction:* Vaclav Smil, "Global Population and the Nitrogen Cycle," *Scientific American* 277, no. 1 (July 1997): 76–81, https://doi.org/10.1038/scientificamerican0797-76.

145 *Some experts fear that humans are now precipitating:* See Elizabeth Kolbert, *The Sixth Extinction: An Unnatural History* (New York: Henry Holt and Company, 2014).

146 *One group of researchers has calculated:* James Mulligan et al., "Carbonshot: Federal Policy Options for Carbon Removal in the United States" (Washington, DC: World Resources Institute, January 2020), www.wri.org/publication/carbonshot-federal-policy- options-for-carbon-removal-in-the-united-states.

149 *Poor people living without air-conditioning are vulnerable in heat waves:* Jan C. Semenza et al., "Heat-Related Deaths During the July 1995 Heat Wave in Chicago," *New England Journal of Medicine* 335, no. 2 (July 11, 1996): 84–90, https://doi.org/10.1056/NEJM199607113350203. Also see Eric Klinenberg, *Heat Wave: A Social Autopsy of Disaster in Chicago,* 2nd edition (Chicago: University of Chicago Press, 2015).

153 *So far, only a few billion dollars a year are being spent on the problem:* COWI, European Commission, and Climate Action DG. Study on EU Financing of REDD+ Related Activities, and Results-Based Payments Pre and Post 2020: Sources, Cost-Effectiveness and Fair Allocation of Incentives, 2018. http://dx.publications.europa.eu/10.2834/687514.

154 *"The forests know how to do this":* Quoted in: Justin Gillis, "Restored Forests Breathe Life into Efforts Against Climate Change," *New York Times,* December 24, 2014, sec. Science, https://www.nytimes.com/2014/12/24/science/earth/restored-forests-are-making-inroads-against-climate-change-.html.

155 *Over those years, global hunger made its largest jump in decades:* According to the Food and Agriculture Organization of the United Nations' The State of Food Insecurity in the World 2009, 1.02 billion people were undernourished worldwide in 2009, more than any time since 1970.

155 *Scientists have calculated that escalating heat:* David Lobell at Stanford University and coauthors found that global corn and wheat production declined by 3.8 and 5.5 percent due to climate change between 1980 and 2008, compared to what yields would have been without climate change. See David B. Lobell, Wolfram Schlenker, and Justin Costa-Roberts. "Climate Trends and Global Crop Production Since 1980." *Science* 333, no. 6042 (July 29, 2011): 616–20. https://doi.org/10.1126/science.1204531.

156 *"What a horrible world it will be if food really becomes short":* Quoted in: Justin Gillis, "A Warming Planet Struggles to Feed Itself," *New York Times*, June 4, 2011, https://www.nytimes.com/2011/06/05/science/earth/05harvest .html.

157 *Where American farmers in the Midwest typically produce:* Figures on global corn yields are published by the United States Department of Agriculture. See Table 04 Corn Area, Yield, and Production in "World Agricultural Production." Circular Series. United States Department of Agriculture, Foreign Agricultural Service, November 2021, https://apps.fas.usda.gov/psdonline /circulars/production.pdf.

159 *After decades of discussion in the United States:* "Gallup Poll Social Series: Consumption Habits" (Gallup News Service, July 2018), secs. 32, 33.

159 *Almost a quarter of Americans told Gallup:* Gallup Inc, "Nearly One in Four in U.S. Have Cut Back on Eating Meat," Gallup.com, January 27, 2020, https:// news.gallup.com/poll/282779/nearly-one-four-cut-back-eating-meat.aspx.

159 *When Herbert Hoover was running for president:* Herbert Hoover, "A Chicken in Every Pot" Political Ad and Rebuttal Article in *New York Times*, Series: Herbert Hoover Papers: Clippings File, 1913—1964, 1928.

159 *Beef consumption per person has actually dropped:* "Food Availability (Per Capita) Data System: Red Meat, Poultry, and Fish" Economic Research Service, U.S. Department of Agriculture, January 5, 2021, https://www.ers.usda .gov/data-products/food-availability-per-capita-data-system/.

160 *For beef cattle, the answer is approximately:* Calculated from Table 3 in Vaclav Smil, "Worldwide Transformation of Diets, Burdens of Meat Production and Opportunities for Novel Food Proteins," *Enzyme and Microbial Technology* 30, no. 3 (March 2002): 308, https://doi.org/10.1016/S0141-0229(01) 00504-X.

160 *One research group at the University of Oxford:* Marco Springmann et al., "Health-Motivated Taxes on Red and Processed Meat: A Modelling Study on Optimal Tax Levels and Associated Health Impacts," *PLOS ONE* 13, no. 11 (November 6, 2018): e0204139, https://doi.org/10.1371/journal.pone.0204139.

Chapter 7: The Stuff We Make

165 *But the great mill in Pueblo, Colorado, has seen its share of misfortune:* Thomas G. Andrews, *Killing for Coal: America's Deadliest Labor War*, paperback edition (Cambridge, Mass: Harvard University Press, 2010.)

167 *In fact, this single industrial process is responsible:* Jeffrey Rissman et al., "Technologies and Policies to Decarbonize Global Industry: Review and Assessment of Mitigation Drivers through 2070," *Applied Energy* 266 (May 2020): 114848 fig. 2, https://doi.org/10.1016/j.apenergy.2020.

170 *Altogether, motors consume half the world's electricity:* "World Energy Outlook 2016" (Paris: International Energy Agency, 2016), 298.

171 *Fat pipes with gentle bends save energy:* Amory Lovins pioneered the concept of system efficiency with coauthors as described, for example, in Paul Hawken, Amory Lovins, and L. Hunter Lovins, *Natural Capitalism: Creating the Next Industrial Revolution,* (New York: Little, Brown and Co, 2000).

173 *And because trading of permits is allowed:* Note that the long-run effectiveness of the California emissions-trading system remains an open question. For a critical review of its features, see Danny Cullenward and David G. Victor, *Making Climate Policy Work* (Cambridge, UK: Polity Press, 2020).

178 *In the United States, 40 percent of concrete and steel:* Ali Hasanbeigi and Harshvardhan Khutal, "Scale of Government Procurement of Carbon-Intensive Materials in the U.S." (Tampa Bay Area, Fla.: Global Efficiency Intelligence, LLC, January 2021), 20.

Chapter 8: Inventing Tomorrow

187 *For decades now, the idea of a coming "hydrogen economy" has shimmered:* For an exemplary specimen of hydrogen hype, see: Jeremy Rifkin, *The Hydrogen Economy: The Creation of the World-Wide Energy Web and the Redistribution of Power on Earth* (New York: *J. P. Tarcher/Putnam, 2002).*

188 *The International Energy Agency recently reported:* International Energy Agency, "*Global Hydrogen Review* 2021."

190 *Electrolyzers have jumped onto the learning curve:* Subramani Krishnan et al., "Power to Gas (H2): Alkaline Electrolysis," in *Technological Learning in the Transition to a Low-Carbon Energy System* (Amsterdam: Elsevier, 2020), 165–87, https://doi.org/10.1016/B978-0-12-818762-3.00010-8.

192 *Within days, 140,000 people had fled central Pensylvania in panic:* A definitive account of the Three Mile Island incident, and the effects on surrounding communities, can be found in J. Samuel Walker. *Three Mile Island: A Nuclear Crisis in Historical Perspective. (*Berkeley: University of California Press, 2004).

192 *Still, the truth is that far more people have died from the burning of fossil fuels:* Researchers from universities in Denmark, the US, and the UK calculated a total of 4,803 fatalities from nuclear accidents between 1950 and 2014. The authors noted that while nuclear accidents are uncommon, the ones that do occur tend to be horrifically expensive. See Benjamin K. Sovacool, Rasmus Andersen, Steven Sorensen, Kenneth Sorensen, Victor Tienda, Arturas Vainorius, Oliver Marc Schirach, and Frans Bjørn-Thygesen. "Balancing Safety with Sustainability: Assessing the Risk of Accidents for Modern Low-Carbon Energy Systems." *Journal of Cleaner Production* 112 (January 2016): 3952–65. https://doi.org/10.1016/j.jclepro.2015.07.059.

193 *But a plant in South Carolina was canceled:* Thad Moore, "How the US Government Wasted $8 Billion and Stranded Tons of Plutonium in South Carolina." *Post and Courier*, cccessed December 1, 2021, https://www.postandcourier.com/news/how-the-us-government-wasted-8-billion-and-stranded-tons-of-plutonium-in-south-carolina/article_24bc000a-da1d-11e9-bb44-87644323c969.html.

199 *Ultimately, the project was canceled:* For the definitive history of the ZeroGen project and its difficulties, see A. J. Garnett, C. R. Greig and M. Oettinger, "Zerogen IGCC with CCS: A Case History," the University of Queensland, 2014.

200 *In 2021, workers finally blew up the part of the plant:* Kristi E. Swartz, "The Kemper Project Just Collapsed. What It Signifies for CCS," accessed December 1, 2021, https://www.eenews.net/articles/the-kemper-project-just-collapsed-what-it-signifies-for-ccs/.

200 *Mississippi, one of the poorest states in the United States, will be saddled:* Steve Wilson, "Two Years since Kemper Clean Coal Project Ended—Mississippi Center for Public Policy," accessed December 1, 2021, https://mspolicy.org/two-years-since-kemper-clean-coal-project-ended/.

202 *People were just getting off work that evening:* The opening anecdote of this section, describing the earthquake in Basel, is based largely on the reporting of James Glanz in the *New York Times*. See James Glanz, "Deep in Bedrock, Clean Energy and Quake Fears." *New York Times*, June 23, 2009.

204 *In a major report in 2006:* Massachusetts Institute of Technology, ed., *The Future of Geothermal Energy: Impact of Enhanced Geothermal Systems (EGS) on the United States in the 21st Century: An Assessment* (Cambridge, Mass.: Massachusetts Institute of Technology, 2006).

204 *One startup company, in Canada:* "Eavor Media Kit," accessed October 8, 2021, https://eavor.com/wp-content/uploads/2021/07/Eavor-Media-Kit-17.pdf.

211 *"If it's wrong to wreck the planet":* Bill McKibben, "Hit Fossil Fuels Where It Hurts—the Bottom Line," *Rolling Stone* (blog), May 21, 2018, https://www.rollingstone.com/politics/politics-news/hit-fossil-fuels-where-it-hurts-the-bottom-line-627746/.

Chapter 9: Saying Yes

213 *"Got our breakfast as early as usual":* James Abbey, *California. A Trip Across the Plains, in the Spring of 1850, Being a Daily Record of Incidents of the Trip . . . and Containing Valuable Information to Emigrants* (Tarrytown, N.Y., reprinted, W. Abbatt, 1933: Library of Congress, 1850), 8, https://www.loc.gov/item/33009652/.

213 *"The wind blew hard enough all night":* Ibid., 10.

214 *The Shoshone tribe of Native Americans reputedly:* Will Bagley, *South Pass: Gateway to a Continent* (Norman: University of Oklahoma Press, 2014), 35.

215 *Bouncing across the dirt roads of the ranch:* The description of the ranch and of Mr. Miller, as well as his quotations, come from a personal tour of the Overland Trail ranch by one of the authors on September 25, 2019.

217 *That fight is about thirty acres of right-of-way:* Sammy Roth, "How a Federal Agency is Blocking America's Largest Wind Farm," *Los Angeles Times*, August 5, 2021.

220 *India has also embraced the* LED *lighting revolution:* Ajinkya Shrish Kamat, Radhika Khosla, and Venkatesh Narayanamurti, "Illuminating Homes with LEDs in India: Rapid Market Creation Towards Low-Carbon Technology Transition in a Developing Country," *Energy Research & Social Science* 66 (August 2020): 101488, https://doi.org/10.1016/j.erss.2020.101488.

220 *On a smaller scale, the clean-power revolution is occurring in Africa:* According to LightingAfrica.org, more than 30 million Africans are meeting their basic electricity needs through off-grid solar products like lanterns.

225 *Private developers have tried to jump-start:* For the best account of an effort to string high-capacity power lines across the nation, see Russell Gold, *Superpower: One Man's Quest to Transform American Energy* (New York: Simon & Schuster, 2019).

235 *For a long time, Mary Anne:* One of the authors of this book, Hal Harvey, is the president of Climate Imperative.

236 *"The science is critical, but it's not enough":* Quoted in Justin Gillis, "For Faithful, Social Justice Goals Demand Action on Environment," *New York Times,* June 20, 2015.

236 *The pope declared that the Earth:* Pope Francis, Laudato Si': On Care for our Common Home, Encyclical Letter, the Vatican, 2015.

238 *Recently, a reporter whose journalism we admire greatly:* Sammy Roth, "California Is Broiling and Burning. Here Are Ideas for Dealing with Climate Despair," *Los Angeles Times,* August 20, 2020, https://www.latimes.com/environment/newsletter/2020-08-20/boiling-point-california-broiling-burning-boiling-point.

Selected Bibliography

Alley, Richard B. *Earth: The Operators' Manual*. 1st edition. New York: W. W. Norton, 2011.

Allwood, Julian, and Jonathan M. Cullen. *Sustainable Materials Without the Hot Air: Making Buildings, Vehicles and Products Efficiently and with Less New Material*. Cambridge, UK: UIT Cambridge Ltd., 2015.

Anadon, Laura Diaz. *Transforming U.S. Energy Innovation*. New York: Cambridge University Press, 2014.

Apt, Jay, and Paulina Jaramillo. *Variable Renewable Energy and the Electricity Grid*. Hilton Park, Abingon, Oxon; New York: RFF Press, 2014.

Archer, David, and Stefan Rahmstorf. *The Climate Crisis: An Introductory Guide to Climate Change*. New York: Cambridge University Press, 2010.

Asmus, Peter. *Reaping the Wind: How Mechanical Wizards, Visionaries, and Profiteers Helped Shape Our Energy Future*. Washington DC: Island Press, 2001.

Bakke, Gretchen Anna. *The Grid: The Fraying Wires between Americans and Our Energy Future*. New York: Bloomsbury USA, 2016.

Bertaud, Alain. *Order without Design: How Markets Shape Cities*. Cambridge, MA: MIT Press, 2018.

Casten, Thomas R. *Turning off the Heat: Why America Must Double Energy Efficiency to Save Money and Reduce Global Warming*. Amherst, NY: Prometheus Books, 1998.

Chase, Jenny. *Solar Power Finance without the Jargon*. New Jersey: World Scientific, 2019.

Clark, Wilson, and David Howell. *Energy for Survival: The Alternative to Extinction*. Anchor Books. Garden City, NY: Anchor Press [u.a.], 1975.

Courland, Robert. *Concrete Planet: The Strange and Fascinating Story of the World's Most Common Man-Made Material*. Amherst, NY: Prometheus Books, 2011.

Cullenward, Danny, and David G. Victor. *Making Climate Policy Work*. Cambridge, UK: Polity Press, 2020.

Dougherty, Conor. *Golden Gates: Fighting for Housing in America*. New York: Penguin Press, 2020.

Dunham-Jones, Ellen, and June Williamson. *Retrofitting Suburbia Case Studies: Urban Design Strategies for Urgent Challenges*. 1st edition. Hoboken, NJ: Wiley, 2020.

Elkind, Ethan N. *Railtown: The Fight for the Los Angeles Metro Rail and the Future of the City*. Berkeley: University of California Press, 2014.

Ewing, Jack. *Faster, Higher, Farther: The Volkswagen Scandal*. 1st edition. New York: W. W . Norton & Company, Independent Publishers Since 1923, 2017.

Fischel, William A and Lincoln Institute of Land Policy. *Zoning Rules!: The Economics of Land Use Regulation*. MA: Lincoln Institute of Land Policy, 2015.

Flannery, Tim F. *Atmosphere of Hope: Searching for Solutions to the Climate Crisis*. New York: Atlantic Monthly Press, 2015.

Foer, Jonathan Safran. *We Are the Weather: Saving the Planet Begins at Breakfast*. 1st edition. New York: Farrar, Straus and Giroux, 2019.

Fogelson, Robert M. *Downtown: Its Rise and Fall*, 1880–1950. New Haven: Yale University Press, 2003.

Ford, Henry. *My Life and Work: An Autobiography of Henry Ford*. United States: Greenbook Publications, 2010.

Fox-Penner, Peter S. Smart Power: Climate Change, the Smart Grid, and the Future of Electric Utilities. Anniversary edition. Washington DC: Island Press, 2014.

———. *Power after Carbon: Building a Clean, Resilient Grid*. Cambridge, MA: Harvard University Press, 2020.

Fraker, Harrison. *The Hidden Potential of Sustainable Neighborhoods: Lessons from Low-Carbon Communities*. Wasington DC: Island Press, 2013.

Freeman, S. David. *Energy: The New Era*. New York: Walker, 1974.

Freeman, S. David, and Leah Y. Parks. *All-Electric America: A Climate Solution and the Hopeful Future*. Place of publication not identified: Solar Flare Press, 2016.

Galbraith, Kate, and Asher Price. *The Great Texas Wind Rush: How George Bush, Ann Richards, and a Bunch of Tinkerers Helped the Oil and Gas State Win the*

Race to Wind Power. 1st edition1. Peter T. Flawn Series in Natural Resources, no. 6. Austin: University of Texas Press, 2013.

Gallagher, Kelly Sims. *The Globalization of Clean Energy Technology: Lessons from China. Urban and Industrial Environments*. Cambridge, MA: MIT Press, 2014.

Gardiner, Stephen Mark. *A Perfect Moral Storm: The Ethical Tragedy of Climate Change*. Environmental Ethics and Science Policy Series. New York: Oxford University Press, 2011.

Gates, Bill. *How to Avoid a Climate Disaster: The Solutions We Have and the Breakthroughs We Need*. 1st edition. New York: Alfred A. Knopf, 2021.

Gipe, Paul. *Wind Energy Comes of Age*. Wiley Series in Sustainable Design. New York: Wiley, 1995.

Glaeser, Edward L. *Triumph of the City: How Our Greatest Invention Makes Us Richer, Smarter, Greener, Healthier, and Happier*. New York, NY: Penguin Books, 2012.

Goddard, Stephen B. *Getting There: The Epic Struggle between Road and Rail in the American Century*. New York: Basic Books, 1994.

Gold, Russell. The Boom: How Fracking Ignited the American Energy Revolution and Changed the World. New York: Simon & Schuster, 2014.

———. Gold, Russell. *Superpower: One Man's Quest to Transform American Energy*. First Simon & Schuster hardcover edition. New York: Simon & Schuster, 2019.

———. The Boom: How Fracking Ignited the American Energy Revolution and Changed the World. New York: Simon & Schuster, 2014.

Gore, Albert. *An Inconvenient Truth: The Planetary Emergency of Global Warming and What We Can Do about It*. Emmaus, PA: Rodale Press, 2006.

———. *Our Choice: A Plan to Solve the Climate Crisis*. Emmaus, PA: Rodale, 2009.

Griffith, Saul. *Electrify: An Optimist's Playbook for Our Clean Energy Future*. Cambridge, MA: MIT Press, 2021.

Grubb, Michael, Jean Charles Hourcade, and Karsten Neuhoff. *Planetary Economics: Energy, Climate Change and the Three Domains of Sustainable Development*. New York: Routledge, 2013.

Grübler, Arnulf, and Charlie Wilson, eds. *Energy Technology Innovation: Learning from Historical Successes and Failures*. New York: Cambridge University Press, 2014.

Gullberg, Anders, and Jonas Eliasson, eds. *Congestion Taxes in City Traffic: Lessons Learnt from the Stockholm Trial*. Lund, Sweden: Nordic Academic Press, 2009.

Harvey, Hal, and Robbie Orvis. *Designing Climate Solutions: A Policy Guide for Low-Carbon Energy*. Wasington DC: Island Press, 2018.

Hawken, Paul, ed. *Drawdown: The Most Comprehensive Plan Ever Proposed to Reverse Global Warming*. New York: Penguin Books, 2017.

Heck, Stefan, Matt Rogers, and Paul Carroll. *Resource Revolution: How to Capture the Biggest Business Opportunity in a Century*. Boston: Houghton Mifflin Harcourt, 2014.

Helm, Dieter. *Burn Out: The Endgame for Fossil Fuels*. New Haven, CT: Yale University Press, 2017.

Hempling, Scott. *Regulating Public Utility Performance: The Law of Market Structure, Pricing and Jurisdiction*. Chicago: ABA, Section of Environment, Energy, and Resources, 2013.

Henson, Robert. *The Thinking Person's Guide to Climate Change*. Boston, MA: American Meteorological Society, 2014.

Herman, Arthur. *Freedom's Forge: How American Business Produced Victory in World War II*. 1st edition. New York: Random House, 2012.

Hirschmann, Kris. *The Kuwaiti Oil Fires. Facts on File Science Library*. New York: Facts on File, 2005.

Hirsh, Richard F. *Power Loss: The Origins of Deregulation and Restructuring in the American Electric Utility System*. Cambridge, MA.: MIT Press, 2001.

Hirt, Sonia. *Zoned in the USA: The Origins and Implications of American Land-Use Regulation*. Ithaca NY; London: Cornell University Press, 2014.

Hone, David. *Why Carbon Pricing Matters*. London: Whitefox, 2015.

Horowitz, Roger. *Putting Meat on the American Table: Taste, Technology, Transformation*. Baltimore, MD: The Johns Hopkins University Press, 2006.

Hughes, Sara. *Repowering Cities: Governing Climate Change Mitigation in New York City, Los Angeles, and Toronto*. Ithaca, NY: Cornell University Press, 2019.

Husain, Tahir. *Kuwaiti Oil Fires: Regional Environmental Perspectives*. 1st edition. Oxford, UK; New York: Pergamon, 1995.

International Energy Agency. *World Energy Outlook 2018*. Paris: IEA, 2018.

Isser, Steve. *Electricity Restructuring in the United States: Markets and Policy from the 1978 Energy Act to the Present*. New York: Cambridge University Press, 2019.

Jackson, Kenneth T. *Crabgrass Frontier: The Suburbanization of the United States*. 26. New York: Oxford University Press, 2006.

Jacobs, Meg. *Panic at the Pump: The Energy Crisis and the Transformation of American Politics in the 1970s*. 1st Edition. New York: Hill and Wang, 2016.

Jahren, Hope. *The Story of More: How We Got to Climate Change and Where to Go from Here*. New York: Vintage Books, 2020.

Johnson, Ayana Elizabeth, and Katharine K. Wilkinson, eds. *All We Can Save: Truth, Courage, & Solutions for the Climate Crisis*. 1st edition. New York: One World, 2020.

Jones, Christopher F. *Routes of Power: Energy and Modern America*. Cambridge, MA; London: Harvard University Press, 2016.

Junginger, Martin, Atse Louwen. *Technological Learning in the Transition to a Low-Carbon Energy System: Conceptual Issues, Empirical Findings, and Use in Energy Modeling*. London: Academic Press, 2020.

Kendall, Henry W., and Steven J. Nadis. *Energy Strategies: Toward a Solar Future: A Report of the Union of Concerned Scientists*. Cambridge, MA: Ballinger Publishing Co., 1980.

Kenworthy, Jeffrey R., Felix B. Laube, and Peter Newman. *An International Sourcebook of Automobile Dependence in Cities, 1960–1990*. Boulder, CO: University Press of Colorado, 1999.

Kiechel, Walter. *The Lords of Strategy: The Secret Intellectual History of the New Corporate World*. Boston, MA: Harvard Business Press, 2010.

Klein, Naomi. *This Changes Everything: Capitalism vs. the Climate*. First Simon & Schuster hardcover edition. New York: Simon & Schuster, 2014.

Kolbert, Elizabeth. *The Sixth Extinction: An Unnatural History*. 1st edition. New York: Henry Holt and Company, 2014.

———. Under a White Sky: *The Nature of the Future*. 1st edition. New York: Crown, 2021.

Koomey, Jon. *Cold Cash, Cool Climate: Science-Based Advice for Ecological Entrepreneurs*. Burlingame, CA: Analytics Press, 2012.

Lamoreaux, Naomi R., Daniel M. G. Raff, and Peter Temin, eds. *Learning by Doing in Markets, Firms, and Countries*. National Bureau of Economic Research Conference Report. Chicago: University of Chicago Press, 1999.

Leonard, Christopher. *Kochland: The Secret History of Koch Industries and Corporate Power in America*. New York: Simon & Schuster, 2019.

Lester, Richard K., and David M. Hart. *Unlocking Energy Innovation: How America Can Build a Low-Cost, Low-Carbon Energy System*. Cambridge, MA: MIT Press, 2012.

Levi, Michael A. *Power Surge: Energy, Opportunity, and the Battle for America's Future*. Oxford, UK: Oxford University Press, 2013.

Levinson, David M., and Kevin J. Krizek. *Planning for Place and Plexus: Metropolitan Land Use and Transport*. New York: Routledge, 2008.

Levy, Barry S., and Jonathan Patz, eds. *Climate Change and Public Health*. Oxford: Oxford University Press, 2015.

Lewis, Tom. *Divided Highways: Building the Interstate Highways, Transforming American Life*. Updated edition. Ithaca, NY: Cornell University Press, 2013.

Liotta, P. H., and James F. Miskel. *The Real Population Bomb: Megacities, Global Security & the Map of the Future*. 1st edition. Washington DC: Potomac Books, 2012.

Lloyd, Jason. *The Rightful Place of Science: Climate Pragmatism*. Consortium for Science, Policy, and Outcomes. Tempe, AZ, and Washington DC: 2017.

Lovins, Amory. *Reinventing Fire: Bold Business Solutions for the New Energy* Era. 1. print. White River Junction, VT: Chelsea Green Publishing, 2011.

Lovins, Amory B., ed. *Least-Cost Energy: Solving the CO2 Problem*. Andover, MA: Brick House Publishing Co., 1982.

Lydon, Mike, Anthony Garcia, and Andres Duany. *Tactical Urbanism: Short-Term Action for Long-Term Change*. Washington DC: Island Press, 2015.

Lyskowski, Roman, and Steve Rice, eds. *The Big One: Hurricane Andrew*. Kansas City, MO: Andrews McMeel Publishing, 1992.

Maegaard, Preben, Anna Krenz, and Wolfgang Palz. *Wind Power for the World: The Rise of Modern Wind Energy*. Pan Stanford Series on Renewable Energy, vol. 2. Singapore: Pan Stanford Publishing, 2013.

Malm, Andreas. *Fossil Capital: The Rise of Steam-Power and the Roots of Global Warming*. London: Verso, 2016.

Mazzucato, Mariana. *The Entrepreneurial State: Debunking Public vs. Private Sector Myths*. New York: PublicAffairs, 2015.

McKibben, Bill. Fa*lter: Has the Human Game Begun to Play Itself Out?* 1st edition. New York: Henry Holt and Company, 2019.

McLean, Bethany. *Saudi America: The Truth about Fracking and How It's Changing the World*. New York: Columbia Global Reports, 2018.

McNally, Robert. C*rude Volatility: The History and Future of Boom-Bust Oil Prices*. Center on Global Energy Policy Series. New York: Columbia University Press, 2017.

Meier, Richard L. *Planning for an Urban World: The Design of Resource-Conserving Cities*. Cambridge, MA: MIT Press, 1974.

Mendez, Michael Anthony. *Climate Change from the Streets: How Conflict and Collaboration Strengthen the Environmental Justice Movement*. New Haven, CT: Yale University Press, 2020.

Mildenberger, Matto. *Carbon Captured: How Business and Labor Control Climate Politics*. American and Comparative Environmental Policy. Cambridge, MA; London, UK: MIT Press, 2020.

Miller, Victoria, and Christopher Schreck. T*he Colorado Fuel and Iron Company. Images of America*. Charleston, SC: Arcadia Publishing, 2018.

Montgomery, Scott L., and Thomas Graham. *Seeing the Light: The Case for Nuclear Power in the 21st Century*. Cambridge, UK; New York: Cambridge University Press, 2017.

Naam, Ramez. T*he Infinite Resource: The Power of Ideas on a Finite Planet*. Hanover, NH: University Press of New England, 2013.

Nall, Clayton. *The Road to Inequality: How the Federal Highway Program Polarized America and Undermined Cities*. Cambridge,UK: Cambridge University Press, 2018.

Nordhaus, Ted, and Michael Shellenberger. *Breakthrough: From the Death of Environmentalism to the Politics of Possibility*. Boston: Houghton Mifflin, 2007.

Nordhaus, William D. *The Climate Casino: Risk, Uncertainty, and Economics for a Warming World*. New Haven, CT: Yale University Press, 2013.

Norton, Peter D. *Fighting Traffic: The Dawn of the Motor Age in the American City*. Inside Technology. Cambridge, MA: MIT Press, 2011.

Orr, David W. *Dangerous Years: Climate Change, the Long Emergency, and the Way Forward*. New Haven, CT; London: Yale University Press, 2016.

Palley, Reese. *Concrete: A Seven-Thousand-Year History*. 1st edition. New York: The Quantuck Lane Press, 2010.

Palz, Wolfgang, ed. *Solar Power for the World: What You Wanted to Know about Photovoltaics*. Pan Stanford Series on Renewable Energy, vol. 4. Singapore: Pan Stanford Publishing, 2014.

———. *The Triumph of the Sun: The Energy of the New Century*. Pan Stanford Series on Renewable Energy, vol. 10. Singapore: Pan Stanford Publishing, 2018.

Partanen, Rauli, Janne M Korhonen, and Partanen, Rauli. *Climate Gamble: Is Anti-Nuclear Activism Endangering Our Future?*, 2017.

Pasqualetti, Martin J., Paul Gipe, and Robert W. Righter, eds. *Wind Power in View: Energy Landscapes in a Crowded World*. Sustainable World Series. San Diego: Academic Press, 2002.

Perlin, John. Let It Shine: The 6,000-Year Story of Solar Energy. Fully revised and Expanded. Novato, CA: New World Library, 2013.

Pollack, H. N. *A World Without Ice*. New York: Avery, 2009.

Pooley, Eric. *The Climate War: True Believers, Power Brokers, and the Fight to Save the Earth*. 1st edition. New York: Hyperion, 2010.

Putnam, Palmer Cosslett. *Power from the Wind*. New York: Van Nostrand Reinhold, 1974.

Rabe, Barry George. *Statehouse and Greenhouse: The Emerging Politics of American Climate Change Policy*. Washington DC: Brookings Institution Press, 2004.

Rhodes, Richard. *Energy: A Human History*. First Simon & Schuster hardcover edition. New York: Simon & Schuster, 2018.

Righter, Robert W. *Wind Energy in America: A History*. Norman, OK: University of Oklahoma Press, 1996.

———. *Windfall: Wind Energy in America Today*. Norman, OK: University of Oklahoma Press, 2011.

Rothstein, Richard. *The Color of Law: A Forgotten History of How Our Government Segregated America*. 1st edition. New York; London: Liveright Publishing Corporation, a division of W. W. Norton & Company, 2017.

Sachs, Jeffrey. *The Age of Sustainable Development*. New York: Columbia University Press, 2015.

Sadik-Khan, Janette, and Seth Solomonow. *Streetfight: Handbook for an Urban Revolution*. New York: Viking, 2016.

Scamehorn, H. Lee. *Pioneer Steelmaker in the West: The Colorado Fuel and Iron Company, 1872–1903*. 1st edition. Boulder, CO: Pruett Publishing Co., 1976.

Schlossberg, Tatiana. *Inconspicuous Consumption: The Environmental Impact You Don't Know You Have*. 1st edition. New York: Grand Central Publishing, 2019.

Shaw, Randy. *Generation Priced Out: Who Gets to Live in the New Urban America*. Oakland, CA: University of California Press, 2018.

Sheller, Mimi. *Mobility Justice: The Politics of Movement in the Age of Extremes*. London; Brooklyn, NY: Verso, 2018.

Shere, Jeremy. *Renewable: The World-Changing Power of Alternative Energy*. 1st edition. New York: St. Martin's Press, 2013.

Shoup, Donald C. *The High Cost of Free Parking*. Chicago: Planners Press, American Planning Association, 2005.

Sivaram, Varun. *Taming the Sun: Innovations to Harness Solar Energy and Power the Planet*. Cambridge, MA: MIT Press, 2018.

Smil, Vaclav. *Energy: A Beginner's Guide*. Beginner's Guides. Oxford, UK: One World, 2009.

———. *Energy in World History*. Essays in World History. Boulder, CO: Westview Press, 1994.

———. *Energy Transitions: History, Requirements, Prospects*. Santa Barbara, CA: Praeger, 2010.

———. *Power Density: A Key to Understanding Energy Sources and Uses*. Cambridge, MA: MIT Press, 2015.

Solar Energy Research Institute, ed. *A New Prosperity, Building a Sustainable Energy Future: The SERI Solar Conservation Study*. Andover, MA: Brick House Publishing, 1981.

Speck, Jeff. *Walkable City: How Downtown Can Save America, One Step at a Time*. First paperback edition. New York: North Point Press, 2013.

Sperling, Daniel, and Deborah Gordon. *Two Billion Cars: Driving Toward Sustainability*. Oxford, UK: Oxford University Press, 2009.

Spieler, Christof. *Trains, Buses, People: An Opinionated Atlas of US Transit*. Washington DC: Island Press, 2018.

Stephens, Mark. *Three Mile Island*. 1st edition. New York: Random House, 1980.

Stern, N. H. *Why Are We Waiting? The Logic, Urgency, and Promise of Tackling Climate Change*. The Lionel Robbins Lectures. Cambridge, MA: MIT Press, 2015.

Stokes, Leah Cardamore. *Short Circuiting Policy: Interest Groups and the Battle over Clean Energy and Climate Policy in the American States*. Studies in Postwar American Political Development. New York: Oxford University Press, 2020.

Stoknes, Per Espen. *What We Think about When We Try Not to Think about Global Warming: Toward a New Psychology of Climate Action*. White River Junction, VT: Chelsea Green Publishing, 2015.

Taylor, Simon. *The Fall and Rise of Nuclear Power in Britain: A History*. Cambridge, UK: UIT Cambridge, 2016.

Teplitz, Charles J. *The Learning Curve Deskbook: A Reference Guide to Theory, Calculations, and Applications*. New York: Quorum Books, 1991.

Thomson, Ross, ed. *Learning and Technological Change*. New York: St. Martin's Press, 1993.

Thunberg, Greta. *No One Is Too Small to Make a Difference*. New York: Penguin Books, 2019.

Wagner, Gernot, and Martin L. Weitzman. *Climate Shock: The Economic Consequences of a Hotter Planet*. Princeton, NJ: Princeton University Press, 2015.

Walker, J. Samuel. *Three Mile Island: A Nuclear Crisis in Historical Perspective*. Berkeley: University of California Press, 2004.

Walker, James Blaine. *Fifty Years of Rapid Transit, 1864–1917. The Rise of Urban America*. New York: Arno Press, 1970.

Webber, Michael E. *Power Trip: The Story of Energy*. 1st edition. New York: Basic Books/Hachette Book Group, 2019.

Wright, Theodore P. *Articles and Addresses of Theodore P. Wright*. In four volumes. Buffalo, NY: Cornell Aeronautical Laboratory, 1961, 1970.

Yergin, Daniel. *The New Map: Energy, Climate, and the Clash of Nations*. 1st edition. New York: Penguin Press, 2020.

Illustration Sources and Credits

ALL DRAWINGS BY RAAZIQ MASUD BROWN

Page 7: Share of CO_2 Emissions from Fossil Fuel Combustion by Sector (US 2019)

These are carbon dioxide emissions only, via "Inventory of U.S. Greenhouse Gas Emissions and Sinks: 1990–2019." United States Environmental Protection Agency, 2021, Table 2-1: Recent Trends in U.S. Greenhouse Gas Emissions and Sinks.

Page 19: The Learning Curve of the Model T

Sales price and delivery data are from the Model T Ford Club of America (www.mtfca.com), inflation-adjusted to 1925 USD for years 1909 to 1925. Cost data are volume-weighted by sales price and production volume across all relevant trim lines (Runabout, Touring, Town Car, Landaulet, Coupe, Torpedo Runabout, Delivery Car, Sedan, Coupelet, Fordor, Fordor Sedan, and Tudor Sedan). Since the Model T used similar components across these trim lines, this chart best captures the learning rate across the entire Model T fleet over relevant years. William J. Abernathy and Kenneth Wayne found similar results in "Limits of the Learning Curve" in the *Harvard Business Review*, September 1974.

Page 42: CO_2 Emissions from Fossil Fuel Combustion

Data from: "Inventory of U.S. Greenhouse Gas Emissions and Sinks: 1990–2019." United States Environmental Protection Agency, 2021, Table 2-1: Recent Trends in U.S. Greenhouse Gas Emissions and Sinks.

Page 45: The Learning Curve for Solar Panels

This chart is based upon OurWorldInData.org, with new data extraction and analysis by Rupert Way, University of Oxford. The 1980 to 2020 data from: the *Fraunhofer Institute for Solar Energy Systems 2021 Photovoltaics Report* (page 46), published July 21, 2021. Fraunhofer's 1980 to 2010 estimation is derived from different sources including Strategies Unlimited, Navigant Consulting, EUPD, pvXchange; from 2011: IHS Markit; Graph: PSE 2021. Pre-1980 data: from The Performance Curve Database at the Santa Fe Institute (http://pcdb.santafe.edu/).

Page 81: LED Light Bulb Price

Historic price adapted from the U.S. Department of Energy, *The Future Arrives for Five Clean Energy Technologies –2016 Update*. The 2021 data found via online search of The Home Depot.

Page 99: New U.S. Car and Truck Fuel Economy

US Environmental Protection Agency. *2021 EPA Automotive Trends Report*. Data available at www.epa.gov/automotive-trends/explore-automotive-trends-data. Accessed September 2021. The 2020 data is preliminary.

Page 133: World Urban and Rural Populations

Historic population data and forecast from: United Nations, Department of Economic and Social Affairs, Population Division (2018). *World Urbanization Prospects: The 2018 Revision*, Online Edition.

Page 151: Converting Land for Human Use

Adapted from: OurWorldInData.org. Forests data from: UN Food and Agriculture Organization (FAO); and Williams, M. (2003). Deforesting the earth: from prehistory to global crisis. Agriculture data post-1950 from UN FAO; pre-1950 data from The History Database of the Global Environment (HYDE).

Page 168: The Scale of Industrial Emissions

The 2014 global energy and process emissions for key industrial sectors: estimated by Jeffrey Rissman, Chris Bataille, Eric Masanet, Nate Aden, William R. Morrow, Nan Zhou, Neal Elliott, et al. "Technologies and policies to decarbonize global industry: Review and assessment of mitigation drivers through 2070." *Applied Energy* 266 (May 2020): 114848. https://doi.org/10.1016/j.apenergy.2020.114848. U.S. Transportation Emissions from U.S. EPA GHG Inventory 2020.

Page 195: Two Charts on Nuclear Power

Source: International Energy Agency (IEA). *Electricity Information*, April 2020 edition.

Page 221: Cost of Electricity from New Power Plants

Price data from Lazard Levelized Cost of Energy Analysis, Version 15.0 (October 2021), Page 8. *Levelized Cost of Energy Comparison—Historical Utility-Scale Generation Comparison*. Design inspired by OurWorldInData.org.

Index

Page numbers in *italics* refer to graphs and charts.

About the Authors

Hal Harvey is the chief executive of Energy Innovation, a nonpartisan climate policy firm that advises leaders in the United States and other countries on how to cut greenhouse gas emissions. An engineer by training, he received his bachelor's and master's degrees from Stanford University. A quarter century before you could buy mass-market electric cars, he built his own, charging it from a solar array.

Justin Gillis is an award-winning journalist with four decades of experience explaining complex issues in understandable language for the readers of major daily newspapers, including the New *York Times*, where he was the lead reporter on climate science, the *Washington Post*, and *Miami Herald*. He was recently a fellow at the Harvard University Center for the Environment.

A Note on the Type

This book was set using Tiempos, a modern typeface originally developed for editorial use. Released in 2010, it is heavily influenced by Plantin and Times, and was designed to be readable, clear, and economical when conveying information in newspapers, magazines, and books.